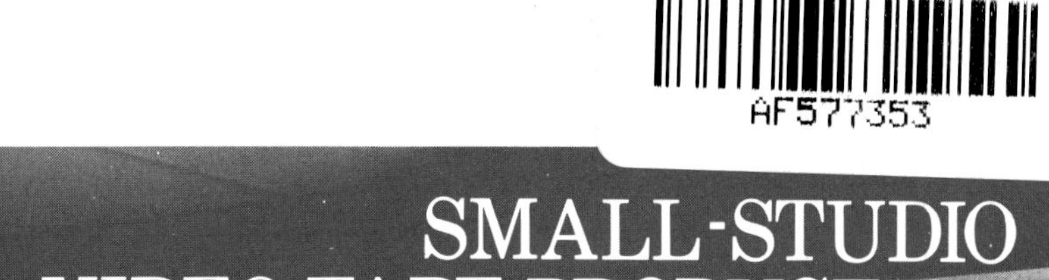

SMALL-STUDIO VIDEO TAPE PRODUCTION

Second Edition

JOHN QUICK
AND
HERBERT WOLFF

ADDISON-WESLEY PUBLISHING COMPANY
Reading, Massachusetts · Menlo Park, California
London · Amsterdam · Don Mills, Ontario · Sydney

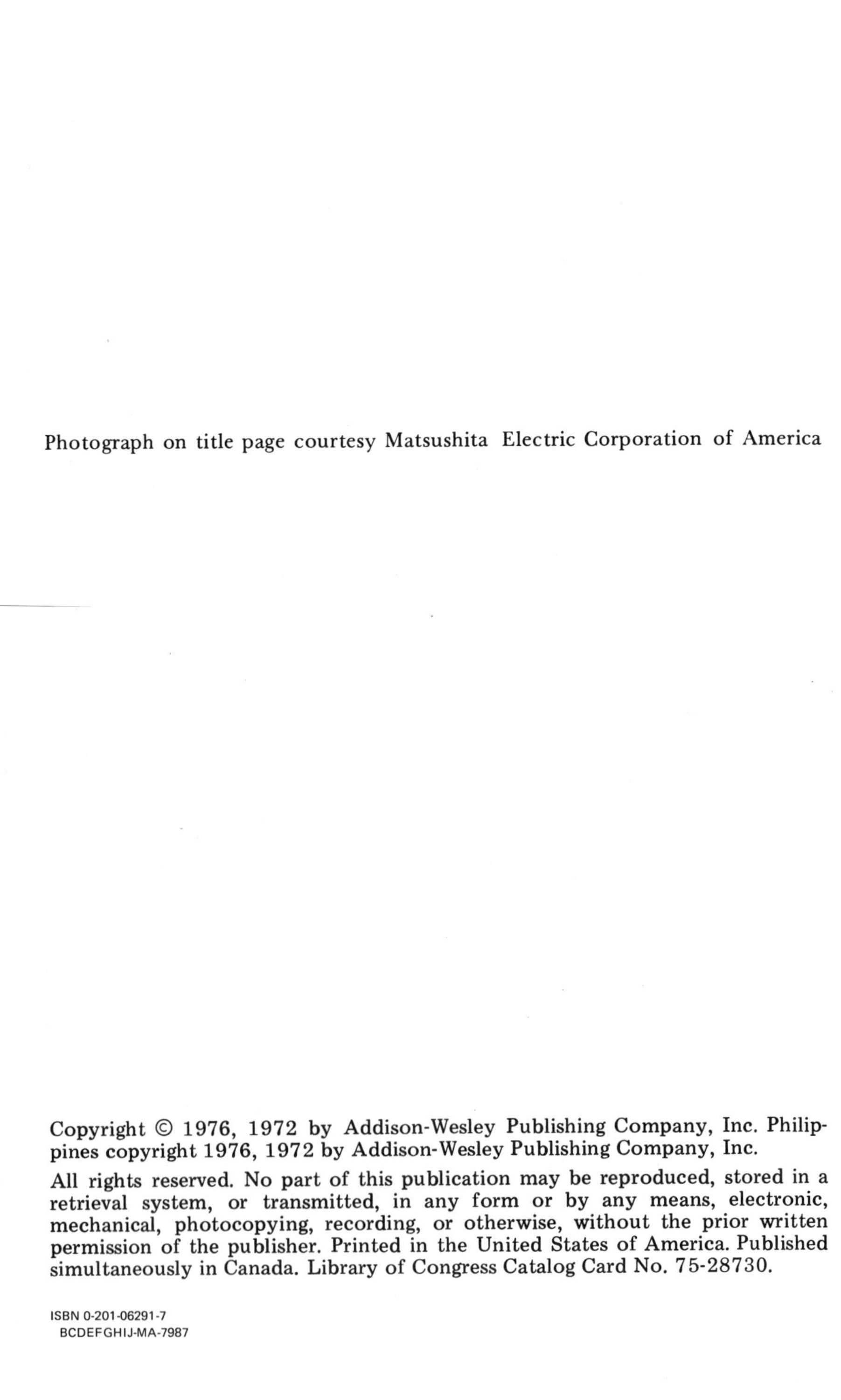

Photograph on title page courtesy Matsushita Electric Corporation of America

 Printed in the United States of America. Published simultaneously in Canada. Library of Congress Catalog Card No. 75-28730.

ISBN 0-201-06291-7
BCDEFGHIJ-MA-7987

PREFACE

Industrial and educational installation of videotape facilities has seen tremendous growth in the past five years. An industry report on life insurance companies, for example, indicated that the number of video tape installations has more than doubled (from 34 in 1970, to over 100 in 1975). Along with the widespread publicity, there have been active marketing efforts behind the sales of video tape and video cassette equipment. You may have purchased some television broadcasting or recording equipment in the past. Or you may be considering the purchase of equipment. At least you have asked the question: "Should I get into video production as a medium of communicating information?" You will probably be correct in answering "Yes."

Within a matter of a few years, probably less than a decade, many communications now being written will be transmitted by media that do not involve writing. Television may well dominate as *the* information system — in industry, in government, and in education. Its impact on marketing alone will be fantastic, since it will enable one to get a sales message into hundreds of selected offices and homes at extremely low cost.

But the medium itself affects the impact of the message, and what appears on the screen will be judged by the same standards as prime-time television. Today's audiences are visually sophisticated. They won't hold still for someone sitting behind a desk speaking into the camera lens. They expect more . . . they're entitled to more.

This is not to suggest that the visual techniques of television are more important than the information being presented. After all, the primary purpose of the small-studio production is to communicate information, not to present entertainment. But information can be presented in an interesting, and often entertaining, manner.

We hope this book will help you in the selection of suitable equipment and other materials. We also hope we can impart information about the organization and layout of a small studio, the kinds of production people necessary to do the work, and finally, some ways to effectively (and enjoyably) produce television programming.

January 1976

J.Q.
H.W.

CONTENTS

CHAPTER 1

THE USE OF TELEVISION AS A COMMUNICATIONS AND EDUCATIONAL TOOL

In 1959, a survey was taken to determine the primary sources of world and local news for most adults in the United States. At that time, newspapers were cited by 54% of the population as their primary source of information, and the most "believable" source of information. Television ranked second to newspapers in 1959, with radio and magazines ranking third and fourth. In 1971, the survey was repeated. This time, 64% of the population cited television as their primary source of world and local news, and the most "believable" source of information. Newspapers now ranked second. Further, of those persons interviewed, over 30% stated they were getting their news *only* from television.

The survey was designed to encourage companies with consumer products to place their advertising dollars in television, and the percentages could be exaggerated depending upon how the questions were asked. But the survey does confirm that since 1948 almost two generations have been brought up to believe in television as their basic communications medium. There are now over 65 million United States homes equipped with over 115 million television sets. Almost 70% of the households report owning a color set. There are 953 television stations operating in this country (including 247 noncommercial stations). In addition, cable television systems now number 3200, serving approximately 10 million people.

Not only does our population gain its information from television in terms of day-to-day news, but they also use it as their major source of entertainment, as a system for making decisions about products, and as a basic instrument for learning. It has served as baby-sitter, teacher, and companion. It has greatly influenced their decisions about many things — from products to presidents.

An additional effect of TV watching is that the audiences have become visually sophisticated, and anticipate seeing pleasant images on a television screen. The other communications media, such as

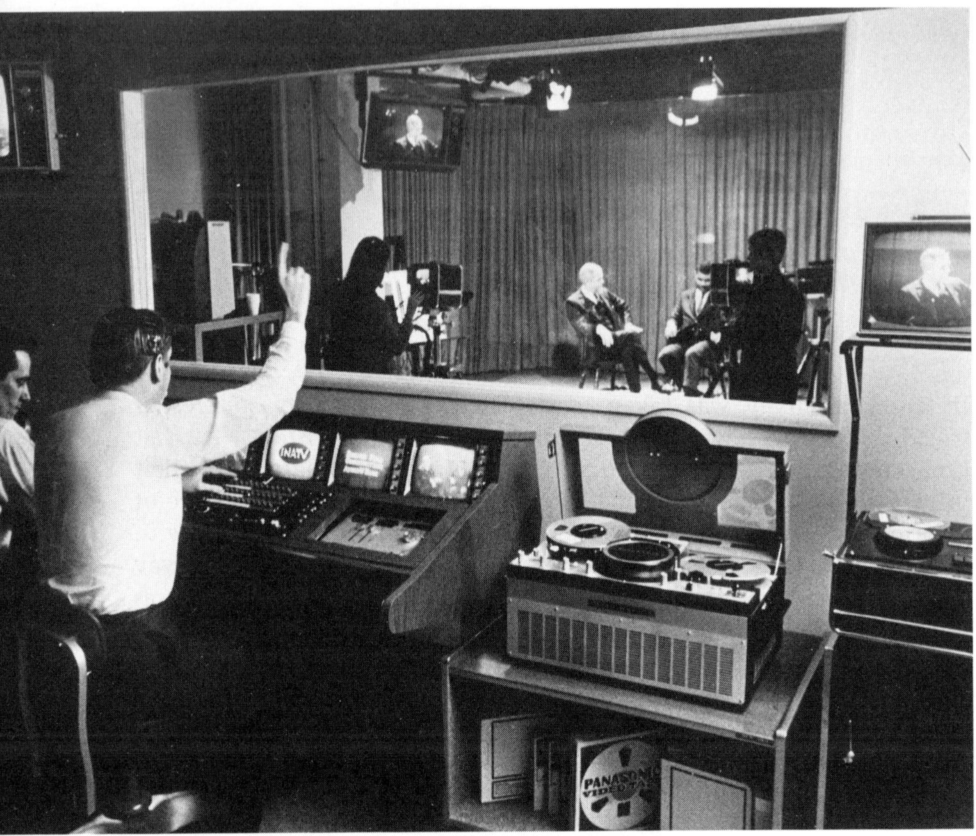

(Photograph courtesy Matsushita Electric Corporation of America)

An in-house television facility offers educational, training, sales, and communications opportunities. Programming content can range from corporate messages to information on new product developments.

newspapers and magazines, have tried to upgrade their visual content in order to approximate the same appeal to general audiences. The ultimate effect is that television has come to set the standards for all communications media. The public now expects its communications to be concise, multi-sensory, visually appealing, and have a sense of truth about them.

If one tries to predict the future of mass communications media — television, motion pictures, magazines, newspapers — it is a reasonable assumption they will have to appeal to segmented audiences. To get a message across effectively to a specified audience,

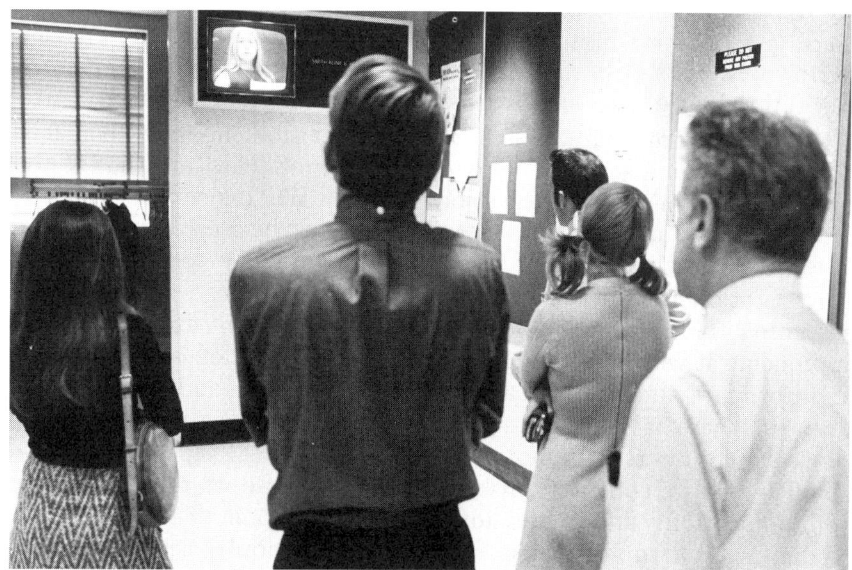

(Photograph courtesy Smith, Kline and French Laboratories, Inc.)

Closed-circuit television being used to supplement employee communications. Note that the TV monitor is in the same location as the employee bulletin board.

communicators will have to be more aware of a group's education, background, and desire for learning. The publications industry has already moved in this direction; motion picture companies and television networks have to follow or face loss of audience.

The industrial and educational user of television must also direct his messages to segmented audiences. The same considerations that determine the use of *any* communications medium must also apply to the use of television. Selection should be on the basis of the ultimate effectiveness of that medium in causing an understanding of the message. Despite the believability of television, it must be kept in mind that it has neither the impact of a "live" presentation nor the concentrated attention and referral capability of a written communication. To record someone's lecture on video tape is not using the medium properly. Thus, our first word of caution: avoid the temptation of using television merely because of its popularity.

The reasons for selecting television (or video tape) as a communications tool include its capability to transmit a sight and sound message over a vast area, and to communicate at less cost than is involved in bringing an audience to a single location or in sending the communicators to various locations.

In all communications, we have to make an assumption that there is a certain willingness on the part of the receiver to consider our signals. This does not imply acceptance by the receiver, merely a certain degree of receptiveness. This assumption of a receptive attitude is more characteristic of the television medium than any other. There is a sense of "immediacy" about television that is not present in other communications media. We expect to see live broadcasting, and even video tape (which may be seen days or weeks later) has a quality comparable to watching a live program. Another consideration for the use of television is that the majority of viewers has come to regard television viewing as a pleasant experience and approaches it with a relaxed feeling. If the information is presented in an informal manner, with care given to style, pace, and visual attractiveness, there is usually a positive reaction to the show and the viewers are more readily motivated.

There is a constant problem about the use of television as a communications and educational tool: it is one of the toughest media. It is a sophisticated integration of people, scripting, electronics, timing, editing, and other factors. When the show is viewed, the audience neither knows nor cares about production conditions or difficulties encountered by the producers. The viewer is used to a certain level of television quality — mainly, the kind he sees on commercial TV. Anything below this level suffers by comparison. (This situation is analogous to professional-versus-home photography. The amateur photographer saves all of his out-of-focus, overexposed pictures, and, despite their viewing faults, the pictures are still of interest — to the photographer. It takes skill and discipline to throw material away, to edit carefully, and to continually strive for better quality.)

A television production, unlike a written document, doesn't have a synopsis or opening summary. The viewer expects to watch the show from beginning to end, and is critical of those parts that don't hold his interest. It is imperative to remember that television is, first of all, a visual medium, and it has the ability to integrate other communications media: people, artwork, charts, slides, photos, film, music, voices, and so forth. But television itself is also a special kind of communications medium — with a special way of viewing the world. This way should be essentially dynamic rather than static. The camera needs to move. The pictures need to change.

In the earliest days of educational television, the producers thought that the *content* of the programming was sufficient to make up for any deficiencies in production. So they permitted a lone camera to stare at a solitary instructor while he lectured. The picture never changed. The camera position didn't change. And audiences went to sleep (if not to another channel).

Since that time, of course, the medium has been applied in a variety of ways, changes have been made (in technique and equipment), and much valuable information and experience have been acquired that will affect the content and the style of contemporary production. At this point let us examine the various familiar and unfamiliar forms of television.

The most familiar, of course, is commercial broadcast television. It implies either network or local programming and uses either live or filmed telecasts, or programming recorded on 2-inch high-band video tape (which gives the same illusion as live broadcasting). The most sophisticated equipment is used, with even small local station facilities requiring the investment of hundreds of thousands of dollars in equipment and facilities (not to mention skilled personnel).

Community antenna television (CATV) refers to a system in which a large, central receiving antenna picks up signals from commercial TV broadcasters and sends them, via cable, to patrons in that community. In some localities, TV viewing would be impossible without a CATV operation. In other areas it merely improves the picture and sound quality to a point where it becomes desirable to buy the service. Operators of CATV systems may also originate their own programming on channels that are not occupied by network or other regularly scheduled programs.

Closed circuit television (CCTV) refers to a system in which one or more television sets are connected to a transmitting cable that carries a signal from a central studio. The signal can only be received by those sets connected to the cable. CCTV includes transmissions within schools, hospitals, etc., and is also used for surveillance systems.

Video tape recording (VTR) was introduced in 1956 and was a vast improvement over the kinescope recordings of the time, kinescopes being motion pictures taken from TV screens. Video tape retained the "live" quality of broadcasting. It also had the capability of being edited and duplicated with a minimum of delay.

Following are some of the other advantages of television, particularly video tape.

1. It creates the next-best thing to personal contact between the speaker (or performer) and the audience.
2. It is essentially a personal, intimate medium. It can be shown in private on a one-viewer-per-screen system, in contrast to film,

which generally presumes an audience numbering from several or dozens to hundreds.

3. It can be produced and distributed in a relatively short space of time.
4. Video tape has immediate play-back capability so that errors can be corrected, and the quality of the *total* production quickly determined.
5. The message can be repeated as many times as the viewer wishes.
6. With certain equipment, the picture can be. slowed down or stopped altogether for closer study.
7. Television tape can be edited and can be integrated with other video tapes.
8. The video can be erased and reused.

The major disadvantages are:

1. There is a large initial cost for equipment.
2. Technical personnel and maintenance costs are a factor with larger systems.
3. There may be a lack of compatibility among various systems.

Video tape comes in more than one size. The largest size, such as that used by the commercial stations, is 2-inch high-band (or "broadcast quality") tape, also called quad tape. Industrial and small studio systems use 1-inch to 1/2-inch tapes, usually referred to as helical scan tapes.

The latest innovation is the introduction of video cassettes (with some manufacturers using a 3/4-inch tape format). A video cassette is merely video tape packaged more conveniently for play-back.

Since most organizations will be utilizing the nonbroadcast 1-inch, 3/4-inch, or 1/2-inch video tape, our text will center on production in these formats. However, the basic principles of studio design, equipment selection, production techniques, and post-production efforts will also apply to CCTV and CATV facilities that originate programming.

There are a number of primary areas in which video tape can be used as a versatile and potent communications tool.

Here is a summary of the types of programs that can be produced in a small studio for industrial and business organizations.

1. *Sales promotion information* can describe the characteristics and benefits of a new product or service. This is of particular interest

to a sales force, and the programming can be aided by the presence of the actual product or new sales literature.

2. *Advertising and marketing programs* can likewise be explained through the medium of television, not only for the benefit of the marketing and sales people directly involved, but also for the information of other personnel in unrelated departments.

3. *Sales training* is particularly adaptable to television, since it enables the viewer to see and hear top sales representatives in actual demonstrations.

4. *Orientation* is the task of every organization, and it must devote a considerable number of man-hours to teaching new employees about the basic benefits and policies of the company. Video tape can cut down this expenditure of time, and remain far fresher and more consistent in its presentation.

5. *Safety* is another responsibility, particularly in manufacturing operations. Continuing safety reports can most effectively be done through the medium of television.

6. *Internal news* programming can be handled through the application of video tape recording — the reporting of new business developments, the introduction of new employees, and adoption of new policies.

7. *Public relations* is an application that can be of interest not only to the employee (more properly employee relations), but to regular visitors to a facility or organization. View-on-demand systems in a waiting room, or special viewings led by an officer of the organization can be easily produced, thus saving lengthy tours (with resultant down-time, danger to the visitor, or disruption of schedules).

8. *Management information* to outlying offices can be programmed periodically or at special times. This is an excellent tool for sending current information about plans, policies, progress reports, etc.

9. *General training* programs offer an opportunity for the viewer to hear experts in a given field and to watch actual demonstrations of products and techniques. The uses extend to every organization and activity: hospitals, specialized training institutions, military organizations, other government operations, business, research, manufacturing, maintenance, and service.

10. *Testing* under field or lab conditions permits observations of certain testing operations in which a human observer would be at a disadvantage. This is not a small-studio application, but any test

coverage that can later be incorporated into a larger tape presentation should be considered.

11. *Presentations* can be rehearsed for analysis and speaker training. The entire presentation can be delivered without interruption so the speaker can achieve a sense of unity and timing. The television camera is kept on the participant so that he can later view the tape and overcome certain of the presentation's problems (such as integrating visual aids and calling attention to physical habits that detract from the overall effectiveness of the presentation).

Because of its importance, the use of video tape in education and training deserves special emphasis.

Television is effective as a medium of instruction. This fact has been demonstrated again and again — with learning groups ranging from preschoolers to elderly people in adult education programs. Television can be used efficiently to teach virtually any subject, ranging from the obvious subjects such as typewriting and foreign languages, to chemistry and anatomy. Naturally, any course to be taught entirely with, or in conjunction with, television must be planned and organized efficiently.

As many training directors and teachers have discovered, the one-on-one aspects of television programming (that has been skillfully blended with supplemental workbooks) make it possible for the student or trainee to progress at an individual speed. The faster-paced person will finish more quickly and can return to a job or go on to other course material. The slower learner may take longer but won't suffer the embarrassment of failing to grasp a point in a classroom.

Television has certain natural advantages:

1. It can provide a much wider window on the world. It is possible to see and hear people that one might not normally come in contact with, such as experts, eminent persons, and panel discussion groups.

2. It makes possible on-site learning. It isn't necessary to travel great distances to avail oneself of the learning experience.

3. With television it is possible to attend class in an informal atmosphere. It can be viewed in a relaxed way in casual surroundings.

4. Because of its nature, the time schedules for television instruction are quite different from classroom courses in which the instructor and the students must agree to meet at specified times. Television viewing can be scheduled at the student's convenience.

5. Lessons can be experienced more than once. Portions of lessons can be selected, or the entire program can be shown in its entirety.

6. There are no interruptions, no distractions, and no questions.

One thing must be remembered about the medium: it should not be expected to carry the whole load of teaching. There are some things it cannot provide a substitute for — such as necessary face-to-face teaching or group supervision. When universities offer television courses, it is necessary to build in feed-back possibilities. Students should be able to get in touch with instructors by telephone, through written communications such as papers and tests, or through occasional classroom meetings on campus or at the extension center.

Once again we must encourage a real understanding of the medium of television. When used in education or instruction it is not an easy way out. Television instruction is not a shortcut. A great deal of attention and work is required to master the skills of television teaching. A classroom teacher who has had no previous experience needs to learn about the medium and how it works and how to best exploit it. It takes a great deal of effort and preparation to make a skillful presentation, one that is more than the face of the instructor speaking on the screen.

CASE STUDIES

The Xerox Corporation in Rochester, New York, felt that television offered unique opportunities for training, engineering research, employee relations, and overall exchange of intracompany information. Taped materials are broadcast through Xerox's internal receiving system in classrooms, conference rooms, and company cafeterias. The system has grown to its own private television network with videocassette systems in 200 Xerox offices around the country.

United States Navy. At the present time all United States Navy ships with a crew of 350 or more are utilizing closed circuit color television for shipboard information, training, and entertainment. A CCTV system provides color origination from a central compartment on the ship via cable distribution to ten or more television receivers. The televised signal can come from such sources as 16mm film, 35mm slides, videotape, or a live camera. Training films and tapes cover such topics as first aid, fire fighting, Navy regulations, basic seamanship, rules of the road, and water safety. Other materials and live lectures relate to specific occupations and tasks on board ship.

The Huntington Public Library in Huntington, New York, has demonstrated that a library can be something more than just a repository of information resurrected from time to time — it can be an active communications center for the creation and dissemination of community information. Videocassettes can be checked out and viewed, in the same manner one checks out a book.

The Corning Glass Works has established a corporate communication network using videocassette equipment. One application of the system is to provide a way for engineers in remote locations to earn a master's degree. The educational material has been made available through Cornell University.

New England Mutual Life Insurance Company has a videocassette network which reaches 119 general agencies throughout the country, stressing field training and communications. It also produces management training tapes, as well as a bi-weekly communication tape viewed by home office personnel.

The Holiday Inn Corporation has a training center which teaches hostelry to over 5,000 managerial employees every year. Two closed circuit TV channels are employed, which are capable of broadcasting to 240 different locations in the training complex.

The Fibreboard Corporation uses TV for training and safety operations and to show portions of the plant under actual working conditions. Tapes are then taken to safety meetings where employees can view hazardous or unsafe conditions that they can correct. Tapes are then sent to other plants.

University and high school applications include: anatomy labs, science labs, psychology classrooms, sociological experimental sessions, drafting lessons, music classes and language instruction, and chemistry.

The medical school applications of video tapes are particularly impressive. Previously only two students at a time were able to study certain dissection procedures in process. With video tape, groups of 10 to 12 can be accommodated. In other areas, one professor with a TV microscope can instruct 150 to 200 students effectively.

Training and educational curricula can be developed (or augmented) through the use of readily available materials from The Public Television Library, a department of the Public Broadcasting Service. Their catalog contains listings for over 1,000 program hours of programming on videocassettes produced by public television stations across the United States. Educators, librarians, colleges, citizens groups, and organizations can use this programming material to serve informational, educational, and instructional purposes — with programs designed to meet very special interests as well as broad general concerns. Subject categories include the arts, environment, health, public affairs and social issues, cities and towns, sports, homemaking and consumerism, and many others.

With more widespread use of television and the opportunity to use small-studio facilities for the production of materials for low-cost playback systems (such as cassettes and other systems described later), more applications will be possible.

Once convinced that television will solve a particular communications or training problem for you, a judgment must be made regarding its practicality from a cost standpoint. We have already implied that to enter into video tape production can be expensive. It is certainly more expensive than the mere purchase price of the equipment. Choices must be made: construction of a studio and purchase of equipment; leasing of components; or the use of outside facilities — all must undergo strict financial consideration. The decision will eventually come down to the question of the cost per show or cost per viewer.

Unless you can anticipate the sustained use of video tape equipment, it is probably best not to construct an in-house studio. By sustained use, we mean a minimum of 50 shows per year for a period of two years.

For example, consider the following costs:

Basic color studio equipment	$70,000
Installation and facility construction	10,000
	$80,000
Divided by two years:	$40,000 per year
Personnel:	
Producer/Director	17,000 per year
Technician	13,000 per year
Production Coordinator	10,000 per year
Total Budget	$80,000 per year

The above figures, of course, do not include salaries of part-time people who may run the cameras, the cost of the studio space, or any maintenance costs (which have to be figured at about 10% per year of your investment costs). You immediately see that 50 shows would cost almost $2,000 each, and the cost per show increases with less volume because all the above costs are fixed. (This may seem like an excessive cost until you realize that the average price these days of *one minute* of a 16mm sound/color motion picture is from $1,500.00 to $2,000.00.)

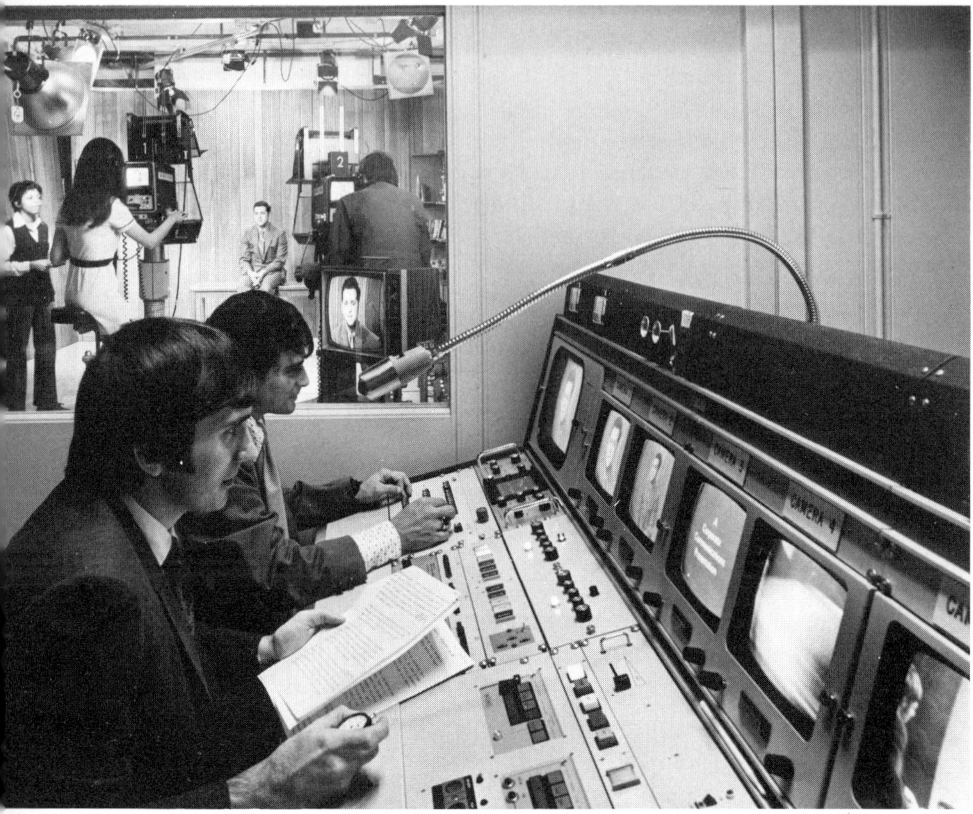

(Photograph courtesy Massachusetts Mutual Life Insurance Company)

Above: A typical small-studio television production facility. Below: Control room of major industrial video tape and CCTV production center.

(Photograph courtesy Merrill Lynch, Pierce, Fenner & Smith, Inc.)

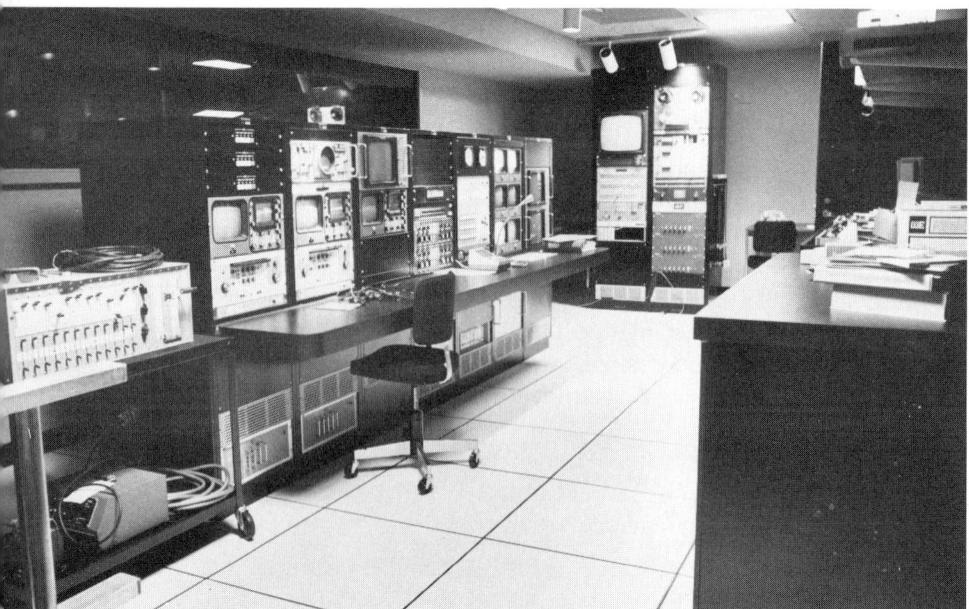

If you contemplate producing fewer than 50 shows per year, it would be wise to investigate the alternative of using outside professional facilities such as a television production house or a local broadcast television studio. If you maintain your outside costs under $1,200 per show, your budget for 50 shows would be approximately the same as if produced in-house, and you would have the advantage of professional production personnel. (A complete description of how to use outside production facilities is in Chapter 16.)

If you plan to spend less than $50,000 to $75,000 for color video equipment (or less than $20,000 for black-and-white), you have automatically precluded an internal television studio per se. For any amount less than that, you are talking about portable-type equipment that will be used for very simple productions. There is certainly nothing wrong with this approach, as long as you recognize the production limitations and lack of in-house post-production refinements.

There is, inevitably, the question: Why not black-and-white? The answer is that new playback equipment is capable of accepting color programs; the additional cost of color studio equipment (especially when amortized over a five-year period) should not be the deciding factor. The basic question is whether to get into television production or not. Having once decided to take the step, the impact of color over black-and-white should convince the purchaser that only in the rarest of instances should black-and-white equipment be considered.

The important point is that the extent of the investment must be determined by the frequency of production and the quality demanded by the audience. One of the first steps, therefore, in determining whether to enter the video tape production field — and to what financial extent — is to list the anticipated productions and how much you are willing to spend for each.

One way of doing this job is to make the best estimates you can regarding video costs, and then compare them with the costs of producing other communications materials (such as training manuals, reports, motion pictures, brochures, and so forth) or developing presentations and conducting regional meetings. You will find that these things are quite expensive, and you may discover some important trade-offs that would lead you to video tape production.

In those situations where cost and level of effort are anywhere near equal, television, with its powerful visual appeal, will inevitably emerge as the most sensible tool.

CHAPTER 2

PROGRAM CONTENT

After publication of his book, *Understanding Media*, in 1964, Marshall McLuhan became one of the most controversial social critics of the time. As author of the famous phrase "The medium is the message," he generated intense praise and equally intense criticism for his "hot" and "cold" media. But, according to McLuhan, he is not concerned with agreement. His real purpose is that of a "neutral probe" — to prod us into rethinking about what it is we are doing, and the environment in which we are doing it. Different environments do, indeed, produce different attitudes and different behaviors. Thus, there can be little disagreement when McLuhan states that the medium "shapes and controls the scale and form of human association and action." Most of us would agree that television has produced profound changes in our society. Also, our attitude *toward* information has changed radically. A communications program, therefore, must be considered in a different light if it is to be produced in the television medium. One can't merely substitute the television screen for the printed or spoken word.

McLuhan states (and here he promotes his greatest controversy) that what is communicated has *less* effect on us than the *means* by which it is communicated. In his book, he insists that the *content* of TV is irrelevant to an understanding of the effects of TV.

It is difficult for us to agree with McLuhan that *what* people watch on TV is as irrelevant as he contends. We believe that certain programming is more effective than others in conveying information and changing attitudes. But we have to agree that the long-term effect of introducing television into a communications system should be a more important consideration than what programs will be produced.

Can a person be taught by television? No question about it. Hundreds of tests have been conducted on children and adults, and the results show conclusively that information can be presented and easily absorbed by television alone or television as a supplementary educational tool.

Although we speak of television as a visual medium, it is obviously more than that. It reaches the audience through two senses, and impressions received from television are carried over and related to other communications. Therefore, certain basic principles must apply to the use of television in the same manner they apply to any communications medium.

Specifically, we are referring to such basic inquiries as audience analysis (prior knowledge, expected action, biases, etc.); timeliness of information; retention and referral requirements; and so forth. In the process of trying to understand the limitations and the expectations of an audience, we must be aware that different kinds of communications are best suited to different techniques. For example, extremely complicated information that needs constant referral (such as a column of figures) obviously requires the printed page. If the enthusiasm and emphasis of a speaker is of prime importance, this may best be communicated by means of an audio tape. If the information would have more impact by demonstration or the visual appearance of a salesman or instructor, then either television or motion pictures should be utilized. If time is an important factor, a written communication would be more effective because it could be produced and delivered with greater speed. If economics is a factor, the message may not be important enough to justify spending the money to make a video tape and send duplicates of the tape to all the audiences involved.

The important point is that just creating communications "stuff," for which there is no definable market, is a waste of time. You will be doing for the sake of doing, not for the sake of communicating for audience understanding and action.

Even after deciding that a video tape or closed-circuit program is clearly the answer to your communications problem, you must then go through an appraisal of the specific audience in regard to a specific medium.

One of the biggest failings of originators of communications is their lack of investigation (and sometimes lack of interest) with regard to their audience. Each audience must be approached in light of its previous experience with the information being offered; the medium by which the information is being transmitted; and the environment in which it is being presented. (A video tape designed to be viewed in someone's home will take on a different tone and pace than one designed to be viewed in a classroom.)

When people have to communicate information, through television or any other means, the first question they usually ask is: "What do I want to communicate to my audience?" Our first suggestion is to amend that question to ask: "What does my audience *need to know*?" The emphasis must shift from *us* to *them*. We often get so involved in the process of telling our story, and hearing the things that *we* want to hear, that we lose sight of the audience and its ability to understand, and eventually to react.

It is important to remember that television is a medium in which the viewer participates. From years of viewing, he has learned to fill in gaps in sales messages. He has learned to make decisions based upon 30- or 60-second messages. He has constructed attitudes on headline news from a 20-second film clip. And in most cases, the viewer prefers it that way. Research has shown that there is a greater sense of participation, identification, and recall with a television message when some of the details have been left to the viewer's imagination.

Therefore, in producing TV, you must keep in mind that it is not a *complete* communications tool. An audience may need prior knowledge or information that supplements the information you intend to give them. Television can't be counted on to do a *complete* training job or a *complete* selling job. For the most part, training and selling are the result of an audience coming in direct contact with a person or a product.

Television, however, can transmit messages, images, interest, and enthusiasm. But as with any communications medium, the transmittal must have an audience objective that is clearly defined. Or stated another way: when a communication is received, what is the intended audience response to the message?

Once you have clearly determined the objective, you are then ready to design a way of getting to it. There are a number of options that are possible:

1. by demonstration
2. by lecture or discussion
3. by example or analogy
4. by dramatic presentation
5. by role-playing
6. all or any combination of the above.

Television has the capacity to encompass many of the optional methods of presentation, and the more ways that are incorporated into the show, the more possibilities of sustaining interest. If a narrator is describing the opening of a new plant, the description will be greatly aided by motion picture footage or still photographs of the facility. Likewise, if a product is being shown and described on a television screen, a similar model can also be in the viewing room for simultaneous reference.

The point is, when planning a television show, do not confine your thinking merely to what a camera in a studio can photograph. Consider first the most desirable approaches, and then see if they have to be modified to fit the limitations of television.

A basic type of television informational program in an industrial or educational organization is the one in which managers, supervisors, administrators, instructors, or marketers wish to communicate with selected groups. This can be achieved either by video tape (to be viewed at a time determined by the viewer) or by closed-circuit television (to be viewed at a time determined by the producer).

A combination of video tape and CCTV is possible. A program can, for example, open with a video tape of an organization's president stating views on a subject of importance. The remainder of the show is presented "live" with other management personnel continuing the discussion and answering questions that are telephoned into the main studio by the audience. Later, video tapes of the program can be shown to other supervisors and line personnel.

Video tape currently has its most extensive use as a training tool — either as the primary training vehicle or as a supplement to a classroom teaching situation. As an instructional tool, video tape can carry the full burden of teaching by a combination of spoken directions and demonstrations. Or, as a supplemental element, it can show a product being used in the field, or tested in a laboratory. Or it can state another viewpoint, or bring experts to the classroom. Simple dramatization of selling situations are another type of training programming. This, of course, is especially helpful where there is a sales "track" — where the story is told in a series of parts and objections are raised and countered within the course of the sales demonstration.

Each organization will find its own special uses for video tape. In the insurance industry, for example, video tape is used effectively to introduce new products to agents located at hundreds of sales offices. Other field communications include presentations by management discussing sales results and introducing new personnel.

Internal or home office video presentations are for role-playing in training sessions; new employee orientation; introduction of new procedures; and so on.

The possibilities for every industry and organization are limitless.

CATV PROGRAMMING

Another application of small-studio production is local programming by Community Antenna Television (CATV) operators. Numerous programs can be originated without too much technical difficulty. The most obvious example is local news broadcasting. This requires the minimum amount of studio space and is a straightforward type of programming.

The advantage of local originations, even to fairly limited audiences or markets, is that there is a fascination about local events. A community, no matter where located in the United States, will devote a certain portion of its attention to local events, particularly if the news is presented professionally. The major drawback to the average local news operation is the absence of visual coverage of key events. The problem can be solved either by motion picture coverage (and the use of a film chain) or with a lightweight, portable video tape recorder and camera.

One of the things that CATV operations can do at extremely low cost, and with the expectation of good response, is programming that relates to the specific audience of the system. As mentioned, strictly local news is always of interest to the viewer because the program tells of things that affect him or his neighbors directly. The announcement that a rabid dog is loose in a community is of far greater importance to the listener than a news story about another summit conference. The listener will not only understand the first story, he will also do something about it. That also goes for all other types of genuine public service programming which tell about the crop conditions, flooding, windstorms, hail warnings, and so forth.

There are other types of public service programming that can also be genuinely useful. Tie-ins with local county agents, fish and game people, law enforcement officers, volunteer fire departments, volunteer rescue operations — all are areas offering rich prospects for interesting and easy-to-produce programming material.

The major point to remember is that your television programming — whether for education, business, CATV, or whatever — must be a "shared" experience. There must be a desired result structured into the program on the part of the presenters, and there must be an involvement and response on the part of the audience.

The television program can be used most effectively if it supplements other communications media. In a training session, television can be a supporting educational tool for a classroom instructor along with materials or models of the subject being discussed. In a marketing situation, a television show can dramatically extol the virtues of a product, and a company representative can follow up by answering specific questions, comparing the product with its competition, overcoming sales resistance, and closing the transaction.

The sequence in which information is presented should follow accepted communication principles. There should be an *introduction* that creates interest in the subject and lets the audience know why the topic is important or will be of value to them. There is a *purpose* that clearly states the ground to be covered and how this relates to other information or materials with which the audience is familiar.

The major portion of the presentation is the *body*. This is where the information is presented logically and effectively. Information that is being presented for training or educational purposes, and is designed to be stopped or replayed for analysis or practice, can eliminate much repetitive discussion. Information to be shown just once may require periodic summaries of what has been covered and how the units of information relate to each other.

When outlining the body of the program, you should avoid allotting equal time to each point being presented. Emphasis of time and production techniques should center on the main point being presented. Thus, if you are selling a product, don't spend a lot of effort on the introduction of the show and the participants. Spend more time and use various means of displaying the product and its advantages. If the program is a discussion of the solution to a problem, then be certain the participants spend more time discussing the solution than discussing the problem itself.

The *conclusion* of the show, in addition to being a wrap-up, should also make its own point. It must ask for some audience response, usually a request for understanding or action.

How long should the program run? The answer, of course, depends upon the content and the visual interest that can be maintained. Note the word *visual*. That is the key to program length.

Television viewer interest depends mainly on the appeal to the senses rather than the intellect. The pleasant and enthusiastic personality of one performer will hold viewer interest much longer than a dull performer who, in fact, may have more to offer on the subject. If, indeed, your selected performer does not project across the screen with the same personality and interest he projects in person, *don't put him on television*. Have him *write* what he has to communicate. Or put him on television supported by other people and visual aids so the attention is spread.

In any case, you are the only one who can determine how long a show should run. You should also keep in mind that viewers have been oriented to commercial broadcast programming of one hour and one-half hour segments, with breaks in concentration within the full time span. Experience has shown that it is difficult to maintain concentrated viewer interest in a "talk show" beyond 40 minutes, and that assumes an extremely high degree of information interest. A standard interview, a lecture, a management message, an employee news program is best kept below 25 minutes. You should exceed that time only if you can add numerous visual elements rather than relying solely on the performers.

Above all, programming for television should be a pleasant experience for both the participants and the viewers. Audiences approach the television screen with the expectation of being entertained or informed. They should not be disappointed.

CHAPTER 3

SPACE REQUIREMENTS

The first consideration regarding studio space is the extent of the television commitment. Once having decided on the quantity and types of programming, you must then determine what equipment will be required and where all the activity will take place. When designing a studio, it is usually safe to say: *get all the space you can get*. As you progress in production, you will probably find a need for additional areas. And the more space one has, the less necessity for shifting and storing equipment, sets, props, and lights. It is also desirable to have enough space to keep a "permanent" set in place for shows which are periodically produced, such as weekly marketing sessions or training programs that utilize a blackboard and slide projectors. The space not taken up by the permanent set can then be used for one-time productions.

When the design of the studio is approached, it must be kept in mind that the studio space is only one part of the overall area needed. Additional space requirements include a control room with enough area for check-out equipment and a small amount of tape storage, as well as table space for editing tape. The control room should have seating capacity for the operating personnel and such outsiders as technical or information advisors. It should not, however, provide seating for management personnel or visitors. The control room is a work area and observers only serve to interfere or distract attention from the production.

A special viewing room for outsiders should be made part of the production facility. It can also serve as a conference room for production discussions and a waiting area for the talent while the technicians are setting up the equipment and rehearsing their camera shots.

The entire television production area should have a single entrance to it, protected by a reception area. This is a most important part of construction, since it will lessen the possibility of people wandering into the studio during the actual shooting, or carrying on conversations that might carry into the studio. Another requirement is a room in which tape is stored (both unused tape and finished productions), and where tape is prepared for distribution.

THE STUDIO

The minimum acceptable space for the studio itself will vary with the frequency and varying types of productions. In general, however, a room about 25 to 30 feet to a side is quite adequate. The space surrounding the actual playing area is needed to change the positions of the cameras and compensate for the limited range of lenses available on standard small-studio cameras.

It should also be remembered that the studio must contain sets (possibly more than one); easels and other carriers for flat art; a paper or drape background; cameras; camera cable (which must be laid without tension and arranged to make it easy, or for that matter, possible for cameras to move to necessary positions); an occasional light on a stand; microphones and microphone cables; and one or more video monitors. The space must also meet a few other basic criteria. It should, in the first place, be an interior room. An outside room is usually next to a passageway and is likely to pick up employee and work noises from inside, and traffic noise from the street.

If the only available space should happen to be a room facing the outside of the building, then be certain that the room is windowless. It is difficult to deal with the changing character of natural daylight, since it can be bright one moment and (because of clouds) dim the next. The area should be lighted with artificial light which is predictable and controllable.

The ceilings should be high for two principal reasons: to enable the lighting instruments to be mounted properly, and to allow a place for the heat generated from the lights to rise. A height of 12 to 13 feet can be considered a safe minimum.

A considerable amount of light is required to render the best results in television production. For this reason a large number of lights will be needed and they must be mounted far enough away from the sets and the performers to allow the light to spread evenly, and also to keep from cooking the talent. In most cases the lights will be operated from the ceiling, not from stands on the floor (which is often the case in location shooting with motion pictures). The lights

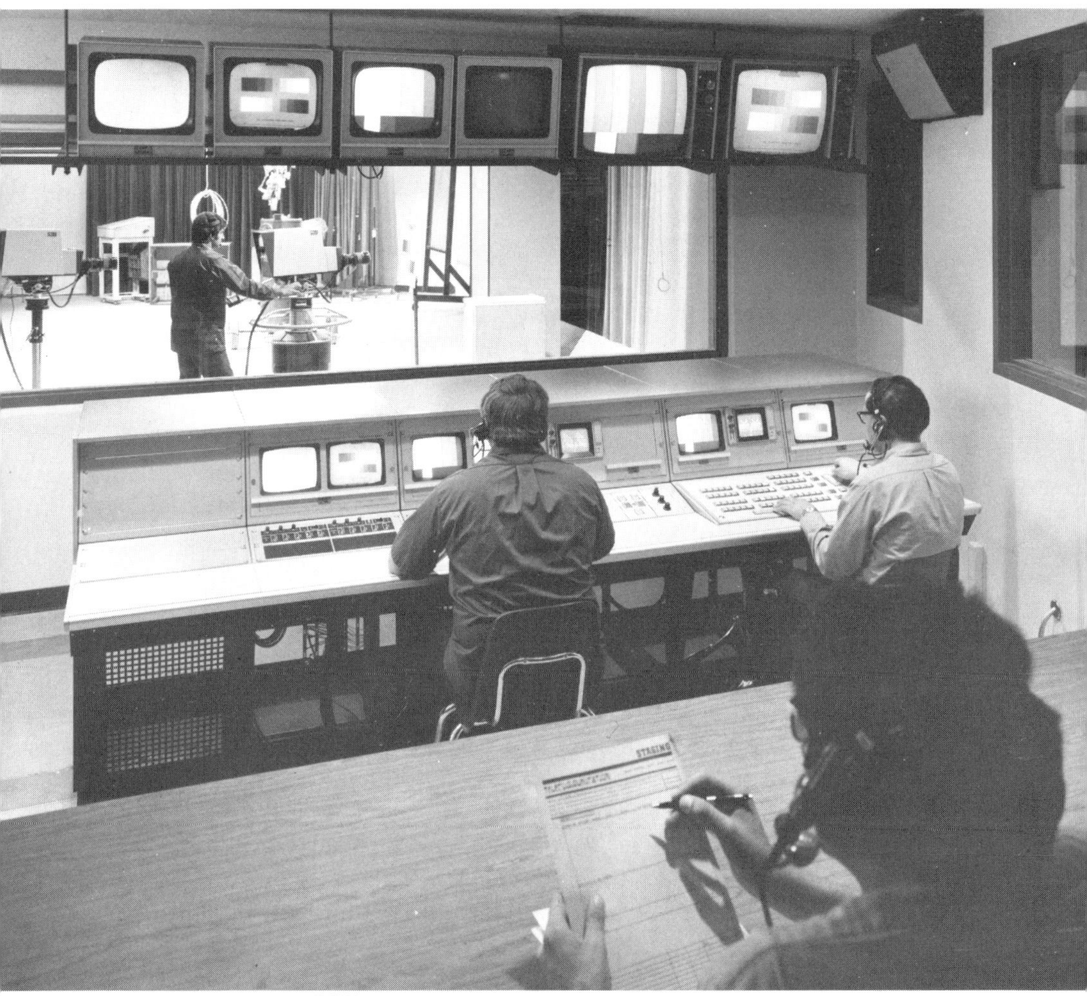

(Photograph courtesy RCA)

TV studio at San Diego City College

for a small TV studio will, for the most part, be at the ceiling on racks or mounts which make it possible to control direction and intensity of the individual units. (Lighting techniques are discussed further in Chapter 12.)

Another important feature is the floor. The flooring material should be as smooth as possible, so that the cameras can be moved easily over it. With most small cameras and dollies it isn't possible to dolly or truck very effectively (that is, produce an effect wherein the camera itself is moving). That generally requires much larger, heavier equipment. The main reason for having a wheeled dolly with a small studio set-up is to make it easy for the camera to be moved from place to place, or from one shot to the next.

Although controlled temperature is a requirement due to the heat generated by the studio lights, one must also be careful of air conditioning noises. The air conditioning ducts should be kept as far away as possible from the performing areas.

The acoustics (or acoustic response) of the small studio or other area in which recording is to be done is an important consideration. There has to be a balance between low and high frequency sounds. Generally speaking, a studio that has smooth, hard, reflective walls and ceilings will allow high frequencies to predominate and will create echo. This condition is known as "hot" or "live." With a certain amount of sound deadening the room will still allow sufficient amount of high frequencies for a pleasant effect. A room dominating in high frequencies is called a "brilliant" room. Too much absorbent material can create a "dead" room.

Most users of small-studio equipment will be doing shows that best lend themselves to table or neck microphones (lavalieres), as opposed to boom or stand microphones. Thus, the problem of acoustics should not be a critical point in determining studio space. However, it may be wise to test the room with a good audio tape recorder. Place the recorder in various parts of the room at medium volume level to see if it picks up any noises that the ear may not distinguish: air conditioning, pedestrian traffic, elevators, typing, etc. Then make sample recordings of voices speaking at different levels and in different areas to determine if wall baffling will be required.

CONTROL BOOTH

In addition to the area required for the studio, there is ideally space for a control booth (or control room). This is the area in which the director (and others, including the technical director, audio man, etc.) controls what is being recorded on the VTR. In the control booth will be found monitors (one for each video source, recorders, or film/slide chain), an air monitor, terminals for the various microphones (so that their sound level can be monitored), audio tape or record input sources, video remote controls for the internal operation of the TV cameras, and switching equipment.

The director must be able to select, and oftentimes mix, various combinations of audio and video inputs. A typical program opening may use a sound effects record or tape, music, and a live or recorded voice while, at the same time, using two live pictures — one superimposed on the other.

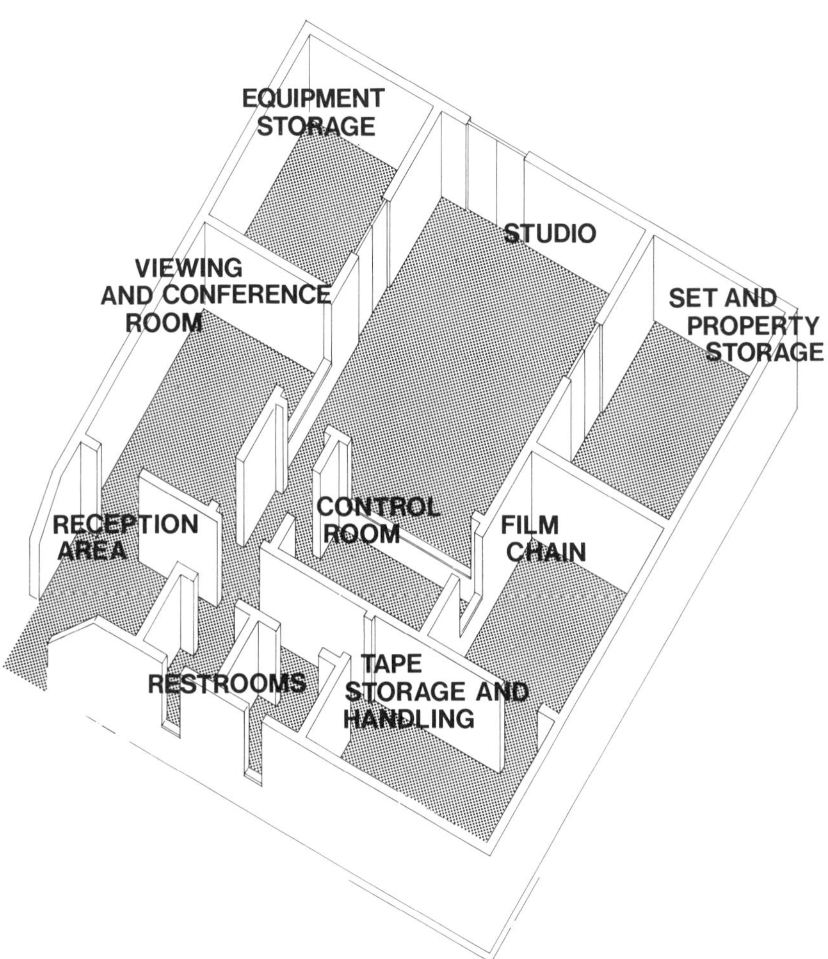

Suggested floor plan for video production facility

In a small-studio situation film chains, if any, should be located in an area next to the control room. (Film chains are used to insert 16mm or Super-8mm motion picture film or 35mm slides into the television format, and hence into the recorded program.)

In the control booth will be located the video tape recorder(s) (VTRs), and there will be the capability to replay the sound and picture of a recorded program in the booth itself, as well as on a separate system to the studio or a special viewing room.

The control room can, technically speaking, be located at some distance from the studio and not necessarily adjacent to it. However, from our experience, it is often a good deal easier to run a show if you are able to see what is going on in the studio from the control booth. For that reason it is desirable to have a double glass window separating the two areas. One of the benefits that accrues is that the director can see not only what is on the monitors, but can, by glancing out to the studio, see the overall scene and easily identify where the camera cables are lying, or where the boom microphone is located, and not call for shots that would require lifting cables or quickly rearranging parts of settings, easels, mikes, etc. This allows potential problems to be spotted before they arise and dealt with more effectively than if the director had only two narrow-vision cameras to depend on.

As we indicated earlier, the control booth is a work area, and access to it must be strictly controlled. There may be required some movement of positions during the taping of a show (such as the engineer moving from the VTR at the opening of the show to the audio mixer to ride sound levels) and the limited space cannot be obstructed by visitors.

Sometimes, however, it is absolutely essential that a nonoperating person be in the booth and seating space for the person must be provided. For example, if a very technical demonstration is being taped, a person from the information group may occasionally have to advise the director where the next center of attention should be. Or if a discussion group is quickly formed for an ad-lib session, the director may need someone to help identify the speakers while the show is in progress. But for the most part, the control booth should contain only the equipment necessary to record the show, and space enough for the production personnel to run the equipment.

STORAGE AREAS

Adjacent to the studio should be two separate storage areas, one for equipment and the other for setting and properties. It is important to separate the two areas for reasons of possible damage and cleanliness.

The equipment storage room should be one that can be properly secured from random access and theft. It will house all the back-up and supplementary video equipment such as extra camera lenses, cables, check-out and maintenance equipment, microphones, extra lights, etc. It should contain cabinets that can be closed and locked as well as high shelf space. If a large inventory of tape must be maintained, this is the place it can be stored.

The set and prop storage area, accessible by double doors from the studio, will be used to store desks, tables, chairs, tripods, blackboards, etc. All extra draperies and paper-roll backgrounds should be kept here. This storage area will also require cabinets to keep small items such as dulling sprays, rags, paints, tools, hardware, etc.

VIEWING AND CONFERENCE ROOM

A combination viewing and conference room is not luxury space — it is nearly essential for an efficient small-studio operation. There is, first of all, a need for a room where discussions can take place in an environment less stressful and distracting than the studio itself. The discussions begin while the studio is being set up, and the director uses the time to brief the participants. Then while the director is taking care of last-minute details, the participants can relax, discuss the questions to be posed, review their notes, and have a cup of coffee. At this time, makeup can also be applied.

The viewing room should be equipped with a special monitor which is fed directly from the control booth console and the control booth VTR. Here is the area in which visitors can watch the show as it is being taped and see exactly what is being recorded.

After the show is taped, there may be a need for an immediate playback to be certain that the main points were adequately covered; that cues were properly responded to; that the recording is free from electronic faults; and just to satisfy the general curiosity of the talent to see how well they did.

Another requirement is for adjacent or nearby restroom facilities. They will be used not only for makeup and primping that is required, but also for performers who are justifiably nervous just before the taping of a program.

ACCESSIBILITY AND CONTROL

It goes without saying that the studio area, particularly, must be accessible from the standpoint of the transportation of things that will be used in programming. For this reason a ground floor location is the obvious choice. Failing that, a location near, but not adjacent to, a large freight elevator is the next best choice.

A part of the accessibility is the need for outside communications to the studio. There must be adequate telephone connections. It is wise to have telephone extensions in the control booth, the

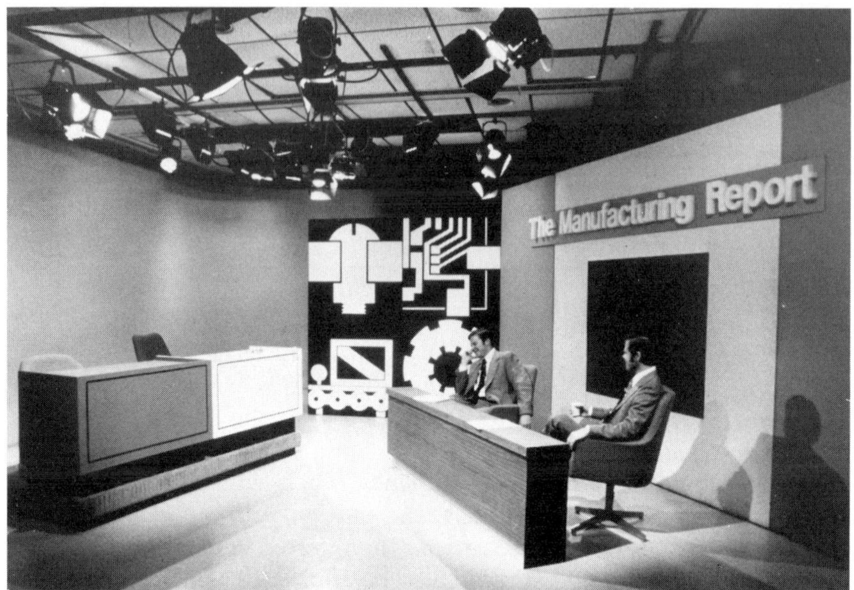

(Photograph courtesy International Business Machines Corp.)

Ample studio space permits flexibility in staging and set design

viewing room, and the studio itself. The telephones should, of course, have a system of flashing lights, rather than bells, to indicate an incoming call.

But the main point is that all accessibility and communication must be controlled. A person familiar with the production operation and the various participants must serve as a control point for traffic into the area, and see that messages and materials are either immediately brought to the proper person or deferred to a more convenient time. A single error in permitting a person to enter the studio at the wrong time, or not immediately delivering materials that are holding up production, can mean time and money wasted, not to mention the frustration and anger these delays can cause.

It goes without saying that the studio should *not* be adjacent to a boiler room or a machine shop, but should be located where outside noise will cause a minimum of problems. (Even the shaft noises of an elevator are easily discernible on a recorded tape.) Noises from the ventilation system and the hum of fluorescent lights should also be thoroughly checked.

It is also a necessity that electrical power be kept constant. Any changes in amperage will be reflected on the video tape and will cause distorted pictures. Therefore, be certain that electrical power lines which feed the television equipment are not on the same circuits as heavy machinery, xerography machines, etc.

Temperature and humidity control is an important consideration for a small-studio facility. In most modern buildings, this will pose no particular problem, since air conditioning and heating are centrally controlled. It may be necessary to calculate a slightly increased need for cooling because of the heat output of the lights.

Humidity control is fairly important, since too dry an environment is conducive to the creation of static electricity. It, in turn, makes the control of dust more of a problem. It must be remembered that the studio and control room have to be kept as clean and dust-free as possible, since the slightest bit of dust on the audio or video recording heads will create problems, either in distortion or attenuation of the signal. Dust in other parts of the equipment causes general kinds of control problems.

The problem of dust raises the question of overall cleanliness in the facility used for producing video recordings. As a general rule, the air must be kept absolutely as clean as possible, and for this reason great care should be taken in what enters the studio or the control room (dirt or dust-carrying materials, etc.).

No foods or beverages should be brought into these areas, and smoking should be forbidden. The area should be swept, dust-mopped, and wet-mopped on a regular schedule, and lighting equipment, sets, furniture, camera equipment, and cables should be wiped down regularly.

There is no excuse for dusty, dirty equipment in a television production facility. It is axiomatic that *any* facility that produces visual end-products should be clean, and it is more than a coincidence that clean printing plants, photographic dark rooms, and movie editing rooms are the ones in which the best work is consistently produced.

CHAPTER 4

EQUIPMENT REQUIREMENTS

Purchasing video tape recording equipment may, indeed, be one of your most frustrating business experiences. Advancements in technology are so rapid, that it is safe to say that the day your equipment is delivered it will probably be outmoded. And having once made the investment, you are then locked into a system that can only be upgraded, and not completely exchanged until you have recovered your full investment by depreciation write-offs or are prepared to step-up to a next generation of systems. However, as is the case with any advancing technological equipment you wish to purchase, you have to start somewhere.

The television industry has not fully worked out, nor is it likely to achieve immediate agreement upon, a common standard of electronic compatibility for nonbroadcast equipment. While compatibility may exist within a single manufacturer's products, at the present time, only limited standardization permits you to record a video tape on one manufacturer's VTR and play it directly back on the VTR of another brand. Most major manufacturers of 1/2-inch black-and-white and color equipment have standardized to an EIA-J (Electronic Industry Association-Japan) format. But, as of this writing, there is no standardization of 1-inch color equipment. In order to play a tape of one manufacturer on the VTR of another, one has to go through an intermediate step and re-record from VTR "A" to VTR "B," and generally in the process lose one degree of quality.

Because of lack of standardization, it is best to approach the purchase of television equipment from the standpoint of a complete system rather than individual components. You must not only determine the system you desire for the studio, but you must also determine where and on what equipment the finished tapes will be played. If your video tapes are strictly for intracompany use, then a

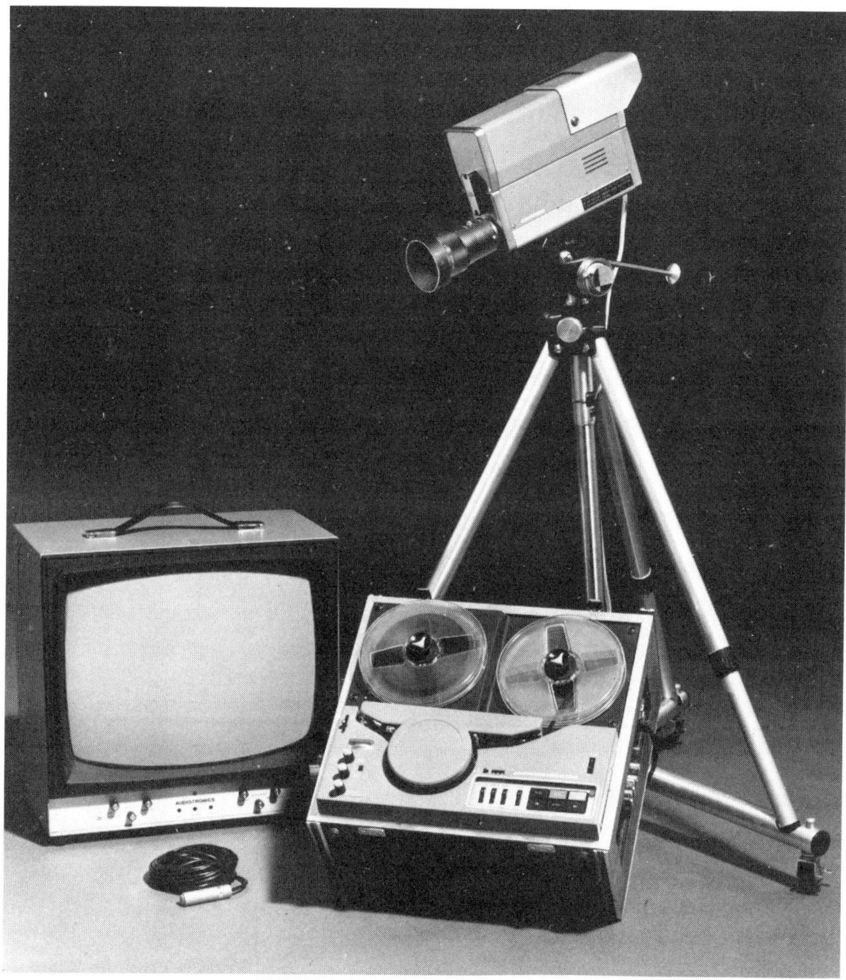

(Photograph courtesy Audiotronics Corporation)

Basic video tape system—video tape recorder, vidicon camera with viewfinder and 30mm lens, TV monitor, and microphone

single manufacturer or distributor may solve your purchasing problem. If you anticipate a wide distribution of your tapes to other companies or organizations using a variety of equipment and formats (widths of tape) then your purchase problems are multiple.

When contemplating the purchase of equipment, you should only consider *color capable* video recording. Technological advances

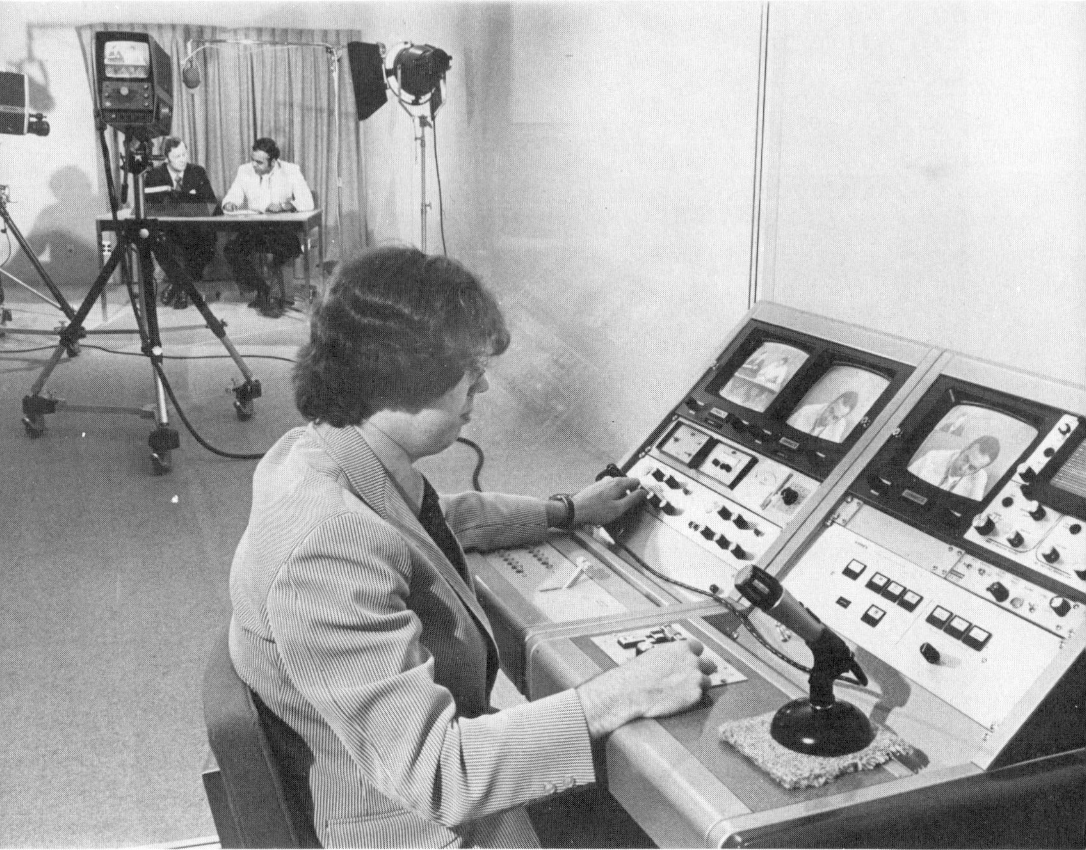

(Photograph courtesy American Republic Insurance Co.)

Small studio and control center

and cost reductions in color television equipment will shortly make black-and-white equipment obsolete. And most of today's equipment, except cameras, is capable of being modified to handle color receiving. However, should you decide your productions need only black-and-white presentation, substantial cost savings can be realized.

The *basic components* required to produce a small studio television show are:

1. camera with tripod
2. video tape recorder
3. monitor
4. audio system
5. lighting

These basic units can be expanded to any degree of sophistication and can include multiple cameras and lenses, a switcher/fader, a synchronizing generator, film and slide chains, extensive lighting equipment, and so forth.

The cost of entering television production can range from a simple hand-held 1/2-inch black-and-white video tape unit priced at under $1200 to complete closed-circuit, multiviewing facilities costing over $500,000.

In addition to facility investment, there is the cost of the production (personnel to produce the shows and operate the equipment), plus the cost of raw tape and completed tapes that form a permanent library.

This chapter sets guidelines for determining the amount of equipment needed for varying degrees of production capability. (The next chapter tells you the personnel needed to plan and execute the program.)

The *basic component* list just cited is merely for a single-camera system. A two- (or more) camera system automatically necessitates multiple components for switching and synchronizing the cameras. The desire to incorporate video tape editing into the productions means the purchase of a second VTR unit, since video tape editing must be done electronically from one VTR to another. And the ability to incorporate films and slides into a show is still another major expense.

We have elected to use the standard of a three-camera 1-inch format studio with editing and film-insertion capability as the basis for our equipment discussion. Such a studio would probably have an average cost of $60,000 for black-and-white recording — $175,000 if it were a color studio.

An important point in purchasing control room equipment is to be certain that it has the capability of handling additional units should you wish to expand your facility. This is especially true with items such as the switcher/fader, which should be able to control a third or fourth camera; or an audio amplifier, which should be capable of controlling five or six microphones.

Quantity	Item
	Studio equipment
2	Vidicon cameras with viewfinders
1	Vidicon camera without viewfinder
2	10:1 zoom lenses F 1.8
2	Heavy duty tripods, heads and dollies (cont.)

Quantity	Item
1	Camera mount
3	50′ camera cables with connectors
1	23″ monitor with stand
5	Lavalier microphones with 50′ cables
5	Intercom headsets
	Control booth equipment
2	1″ video tape recorders
1	Vertical interval switcher
1	Dual-camera control unit
1	Synchronizing generator with gen lock
1	Special effects generator
1	Console rack
4	8″ video monitors
1	23″ monitor
1	Waveform monitor
1 ea.	Audio amplifier and preamp mixer
1 ea.	Audio tape recorder and record turntable
1	Studio audio monitor/intercom system
2	Intercom headsets
	Miscellaneous cables and wiring
1	16mm (or Super-8mm) remote-controlled film projector with stand
2	35mm remote controlled slide projectors with stand/dissolver unit
1	Film camera with remote-control unit
1	Cable assembly
1	Optical multiplexer
	Lighting equipment
8	10″ spotlights — 1000 watt
4	Back lights — 1000 watt
2	Soft lights — 1000 watt
20	Lamps
4	Portable quartz lights with stand
1	Dimmer panel
	Viewing room
1	23″ monitor
1	25′ cable with connectors

COLORIZED TELEVISION AND OTHER COMPLICATIONS

The advent of color brought a new depth of realism to television. It also brought a host of new problems. The increase in enjoyment brought to the viewer was matched by a corresponding increase in cost and complication of television production and transmission.

The basis of the complication is easily explained. The human eye can recognize color errors on the order of 1%, where monochrome errors of 5% or even 10% are relatively tolerable. The end result of this dilemma is to make the accuracy requirements for color transmission roughly ten times greater than monochrome, for an equivalent perceptable quality.

Colorized television is the term chosen to describe the color television system used in the United States. That system is the NTSC (National Television Systems Committee) system, which basically is a monochrome picture with the large areas colored in. The colorization is achieved by encoding (combining) and decoding various quantities of the primary colors red, blue, and green.

VIDEO TAPE RECORDER (VTR)

The underlying principle of magnetic recording, audio or video, is essentially very simple. The sounds and visual images to be recorded are converted to an electric current which varies in intensity corresponding to the level and intensity of the sounds and light to be recorded. This varying electric current is then fed into an electromagnet. Tiny metallic particles (bonded to a tape backing) pass an electromagnet which induces among them a magnetic field which varies in strength according to the variations of the electric current. This forces the metallic particles on the tape to shift position to an extent determined by the varying levels of the magnetic field. Once moved, these particles remain stationary and a recording has been made. When they are again subjected to a current from an electromagnet, the process is repeated and a new recording replaces the old. A single tape may be used for many different recordings in this manner.

To play back an *audio* recording, the tape moves before a stationary head at a set speed. The electromagnetic impulses are then translated into sounds. The uniformity of the tape speed is all that is required to play on machines manufactured by different companies.

In *video* tape recordings, the problem of compatibility arises because both the tape and the video heads reading the tape are moving simultaneously. The speed of the tape, the speed of the head, and the question whether to use a single or double head — these are

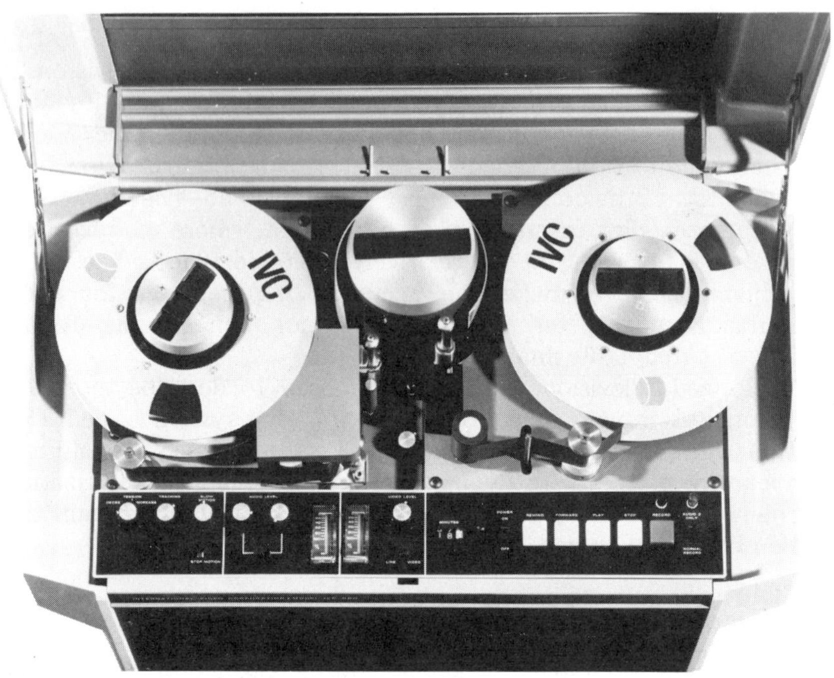

(Photograph courtesy International Video Corp.)

One-inch color helical-scan video tape recorder

decisions made by each individual manufacturer, unless he has agreed to EIA-J standards.

Video tape recorders fall into basic categories based upon the width or format of the tape they record upon: 2-inch, 1-inch, 3/4-inch, 1/2-inch, and even 1/4-inch. In this text we are considering only 1-inch, 3/4-inch, and 1/2-inch formats, since 2-inch tape is basically for commercial broadcasting and 1/4-inch does not have enough versatility or acceptability for industrial or educational use.

Our experience indicates that the 1-inch, 3/4-inch, and 1/2-inch VTRs are of sufficient quality for most users' recording and playback. The degree of quality, of course, varies with the amount invested in the *total* equipment and the skills of the people who operate the equipment.

Most organizations that construct a complete in-house studio and control booth use a 1-inch format as their master recorder and then replicate (duplicate) to the smaller, less expensive 1/2-inch or 3/4-inch format for playback if there is a wide distribution of

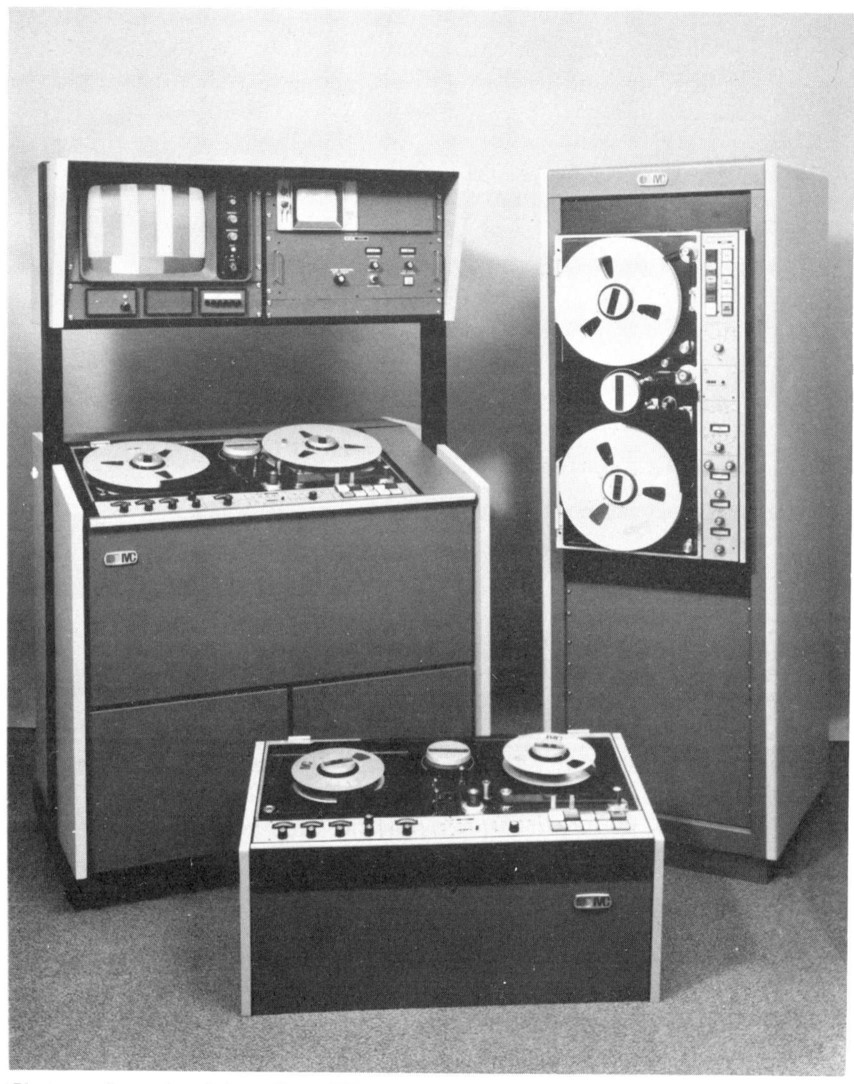

(Photograph courtesy International Video Corp.)

One-inch video tape recorders that range in price from $20,000 to $40,000

audience. If the audience is mainly located in the same facility where the tape was recorded, or if the tape is used for CCTV, then the basic VTR can also serve as the playback unit.

The 1-inch VTR has measurable quality and production advantages over the smaller-width VTR. The broader tape offers a broader

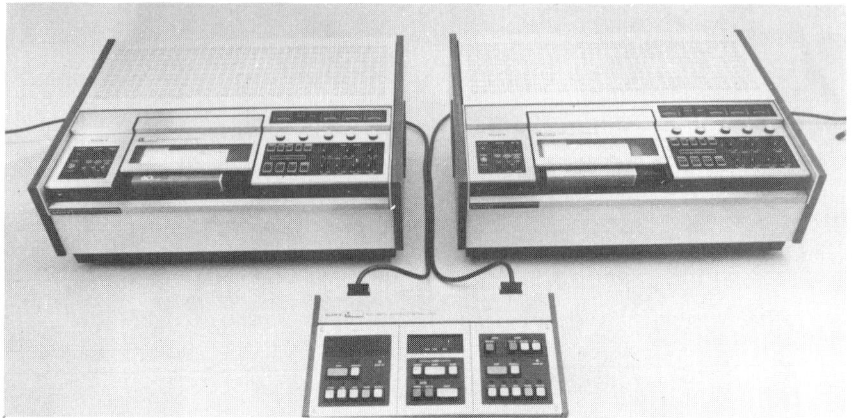

(Photograph courtesy Sony Corporation)

Integrated 3/4-inch video cassette recording and editing system

electronic field and thus has sharper and more stable pictures. The greater electronic stability of the 1-inch format is especially important if numerous duplicates of the program are going to be made. One-inch tape provides easier and more accurate editing capability; and all 1-inch formats provide two audio tracks for later insertion of music or additional narration. In addition, the 1-inch VTR can more readily accept supplementary controls and devices that further refine the audio and video signals.

The portability and lighter weight of the 1/2-inch VTR is certainly more desirable in special situations, such as role-playing in a classroom where there is a need for immediate playback and the quality of the picture is of lesser importance. But unless the finished tape is to be played on the same VTR that recorded it, the 1-inch format is superior for playback and as a master tape for replication purposes.

Recent advances in the recording and editing capabilities of 3/4-inch cassette units have made this format one to be considered for small-studio installation. Its major advantage is that it is an integrated system, and the end product is a cassette master. The major drawback is that the system is less flexible than reel-to-reel systems.

If tape editing and extensive postproduction effort is anticipated, it will be necessary to purchase two VTRs.

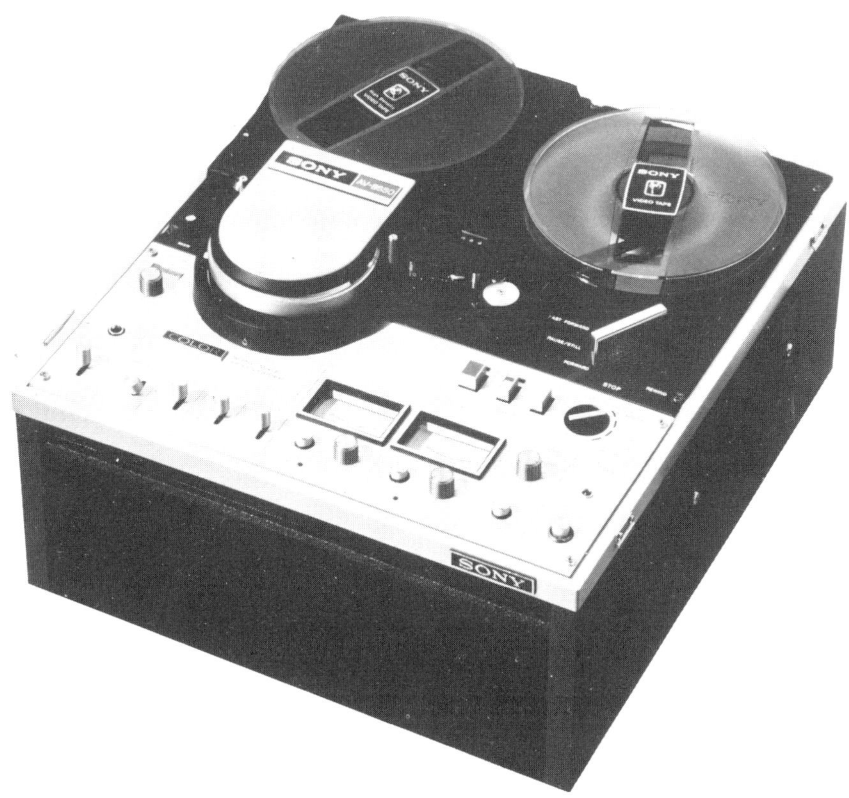

(Photograph courtesy Sony Corporation)

One-half inch helical-scan color video tape recorder with editing capability

It is possible to save a few hundred dollars in VTR costs by purchasing one VTR with editing capability and one without such capability. Normally, an editing machine is externally locked to a synchronizing generator, so that the vertical-interval signals on the editing machine and on the synchronizing generator occur at the same time. If the playback machine feeding the editing machine is not also locked to the same generator, then switching from a "Play" to a "Record" function on the editing machine will cause a relative change of the vertical-interval signal and the new recording will have a vertical roll in the picture. On some editing machines, it is possible to lock the machine to the playback machine. In this case, there will be no vertical roll, but there will be an increase of instability as the machines are not both referenced to another external source.

It is recommended that if extensive editing is anticipated, the increased capability should warrant the purchase of the two larger units. When two complete editing units are used, all transitions can be accommodated without any noticeable disturbance to the picture and the stability of the recording will be improved.

Note: Without the use of external programmers, it is almost impossible, except with luck, to achieve consistent edits better than within 1/2 a second. With some external programmers, editing accuracy within 1/10 of a second can be obtained, and with some recent developments it is possible to do frame-by-frame editing.

TAPE

One of the problems in video tape recording is the oxide buildup on the video heads. (To a much lesser degree, this is also true of audio heads.) Our experience has shown that picture "noise" — snow effect on the screen — is caused by oxide buildup either on the machine that produced the tape or on the playback unit.

It is strongly advised to clean the video heads each time before recording — and be certain that periodic cleaning is maintained on the playback units throughout the communications system.

When selecting the tape upon which to record, find a brand that is not too soft (which causes oxide buildup), and one that is not too hard (which causes extreme wear on the video heads). It is best to stick to brands of video tape produced by "quality" companies.

On the rare occasion when there is a problem with a new tape, it usually manifests itself as insufficient coating or irregularities in the coating. When this condition occurs, there will be a temporary loss or interruption of the picture signal. This condition is called *dropout* and is characterized as a horizontal streak on the television screen. Some video tape recorders may be equipped with dropout compensators, which are components that replace the missing video information with video previously presented.

CAMERAS AND LENSES

The television picture originates when the camera begins to "scan" an image onto a set of horizontal lines. (The greater the number of lines, the sharper the picture.) There is such a thing as purchasing too much camera for the rest of the equipment. Camera selection must be based upon the consideration of the total television system and the desired versatility and quality. All cameras, however, should be

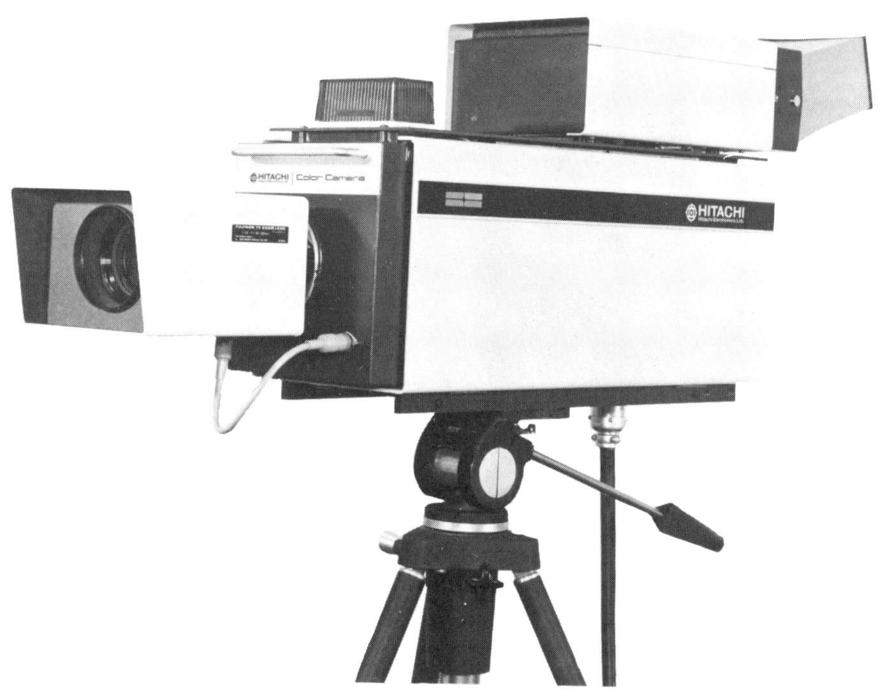

(Photograph courtesy Hitachi Shibaden Corp.)

Broadcast-type three Plumbicon color camera for major industrial and educational studios

of solid-state construction and meet Electronic Industries Association (EIA) standards.

There are four basic types of cameras:

a. image-orthicon — formerly used for commercial telecasting, but no longer manufactured

b. vidicon — used primarily for small-studio production

c. lead-oxide — used for both commercial and small studios. Registered under the trade name Plumbicon by Norelco (Philips North American Co.).

d. silicon-diode — more sensitive and more rugged than a vidicon camera, and used in lower-light situations.

Most industrial color and black-and-white television cameras utilize a vidicon picture tube that produces excellent resolutions under ordinary studio lighting conditions. Horizontal resolution of

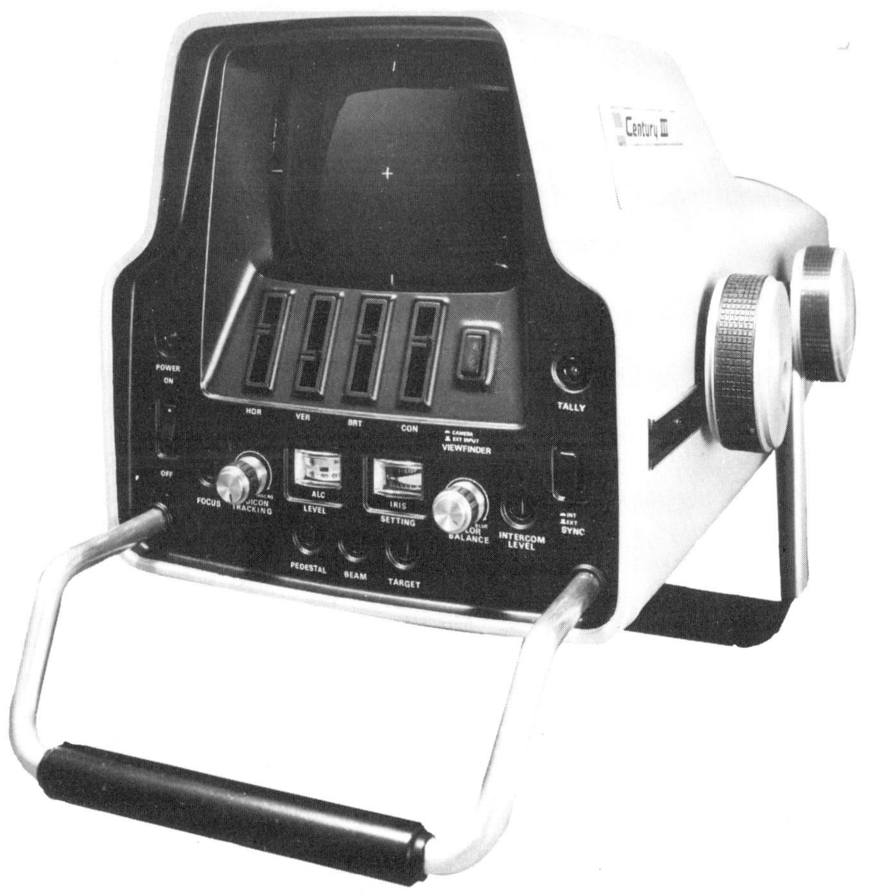

(Photograph courtesy Sharp Electronics Corp.)

Inexpensive single-tube color camera

approximately 550 lines is sufficient for most studio shooting. Where shooting requirements indicate a need for more exact reproduction (such as close-up research projects) then a camera with higher quality horizontal resolution (approximately 750 lines) may be needed.

Two cameras are usually adequate to begin most small-studio productions, especially if a slide chain is included in the overall system. A third camera provides additional flexibility, however, and can assume the role of the slide chain if superimpositions of names, logos, etc., are desirable. A fourth camera is usually considered a luxury, or possibly a back-up unit if downtime is a critical factor in the studio.

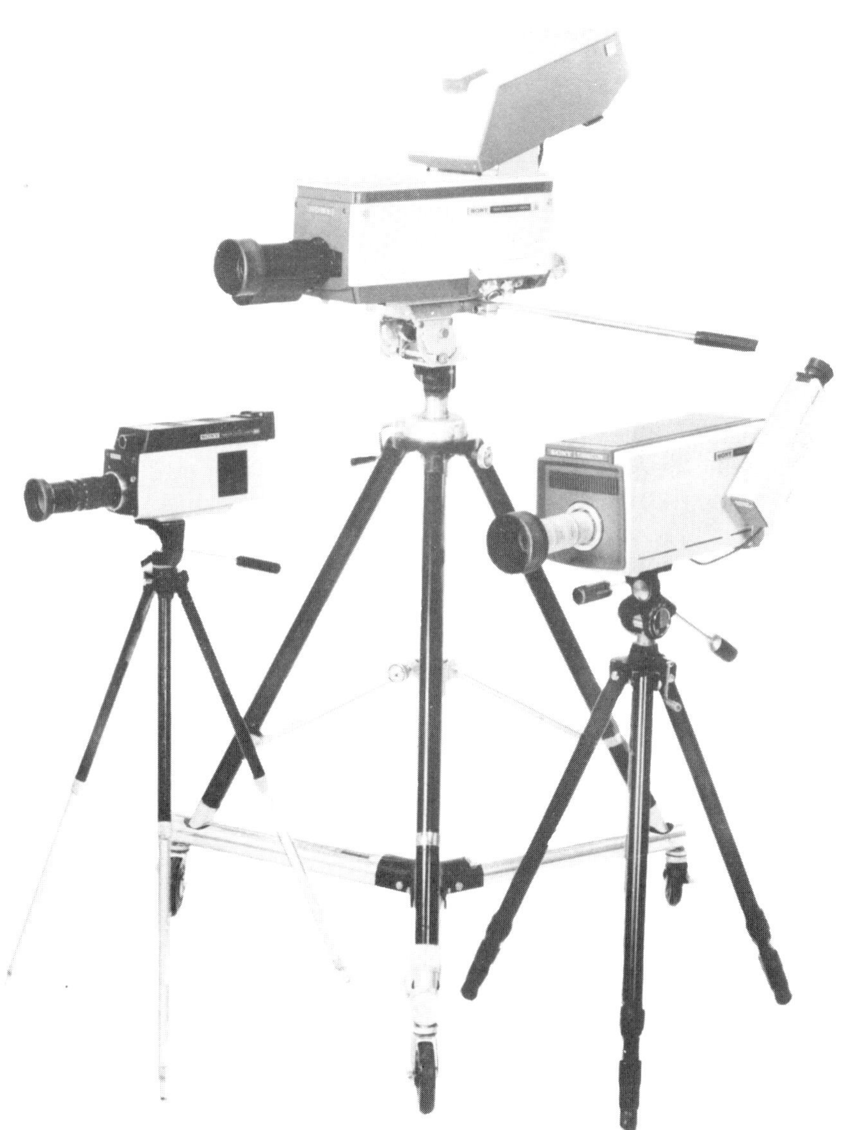

(Photograph courtesy Sony Corporation)

Three single-tube color cameras. The smallest camera is primarily used with battery-operated portable systems.

Most cameras have built-in viewfinders, and cameras of lesser quality are not advised. An important part of any production is that both the director and the cameraman can see what is being recorded.

Cameras should have their own individual VF adjustments for brightness, contrast, and shading, as well as having the capability of being adjusted (for matching purposes) from the control panels in the booth.

Each camera should come equipped with a manually controlled zoom lens. (A zoom lens with a ratio of 4:1 is considered standard, but higher ratios — up to 10:1 — may be required.) If a studio is especially narrow, one camera may require a wide-angle lens to pick up all the participants in a group discussion. However, it is wise to stick with good zoom lenses and position your cameras for wide and close-up shots. Most color cameras are now manufactured with only a single zoom lens. (A more complete discussion of lenses and how they operate will be found in Chapter 9, Camera Technique.)

A characteristic of vidicon cameras is "picture lag." If a person moves quickly in front of the camera, or if the camera "pans" too quickly, the image appears to lag and then catch up to where the person is. The effect lasts for only fractions of seconds and is merely disconcerting. It only serves to remind you that motion by either the performer or the camera must be at a reasonable speed.

The light weight of the camera necessitates a heavy tripod for smooth movement and stability. Although you may not be doing much changing of camera positions during a taping, a solid tripod and dolly can prevent a camera from being knocked over by someone accidentally bumping the unit. And should a friction lock on the tripod head not be set tight, and the camera suddenly swing down, the weight of the tripod will prevent the momentum from pitching the camera to the floor.

If the equipment is purchased and installed by a local distributor or franchisee, he will also align the cameras and balance colors, contrasts, etc. This is necessary since each video system must be adjusted for the lighting conditions under which it will be operating. After the initial settings are made, you should find that only minor adjustments are necessary for day-to-day operations. Should a camera require extensive alignment each time it is used, it is a sign of either a defective or worn-out component that should be replaced. It is always advisable to replace or repair equipment at the first sign of trouble, since a single bad component can cause other components to be overworked and thus cause their early demise.

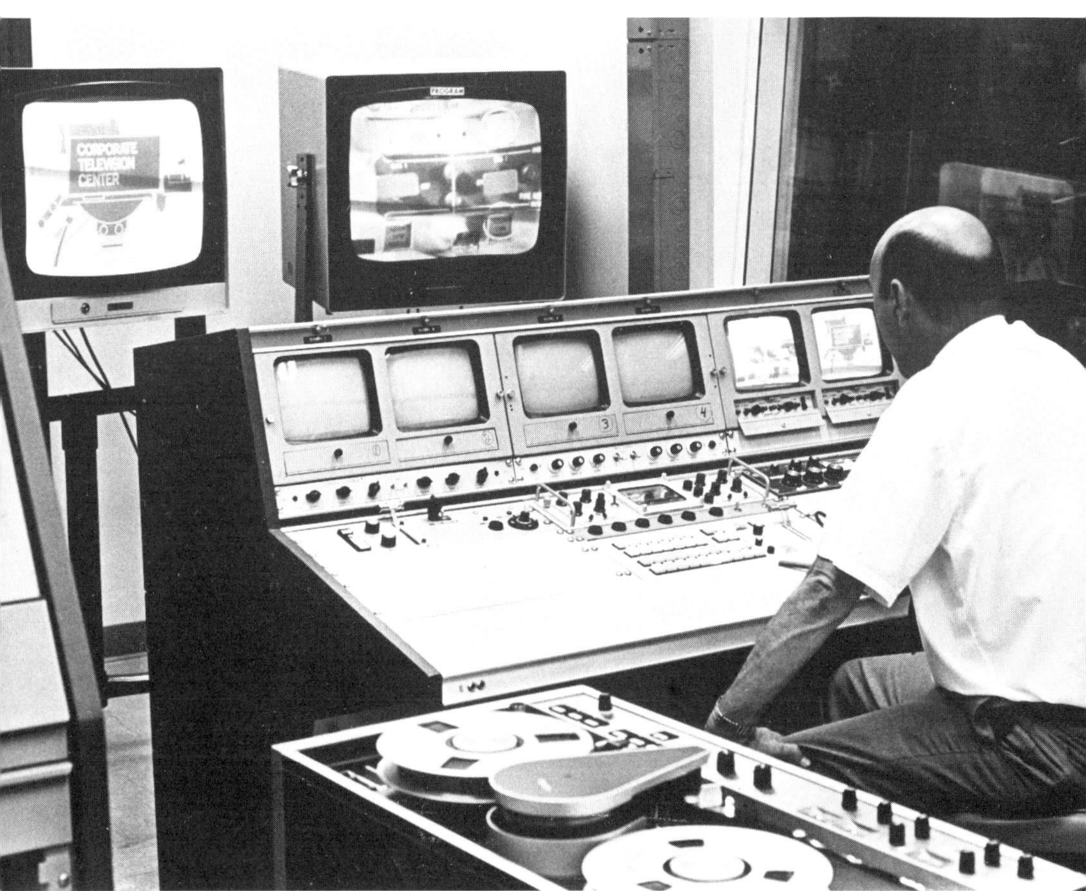

(Photograph courtesy Eastern Airlines)

Switcher panel in control room

SWITCHER/FADER (VERTICAL-INTERVAL PRODUCTION SWITCHER)

A switcher/fader is required when two or more cameras are used. It permits the operator to switch from camera to camera by pressing a button for an instantaneous new picture; or to bring in one image as another image is going out; or to bring in an image from a blank (black) screen or take the picture out to a blank screen. In the vernacular, the three options are called a cut, a dissolve, and a fade-in or fade-out.

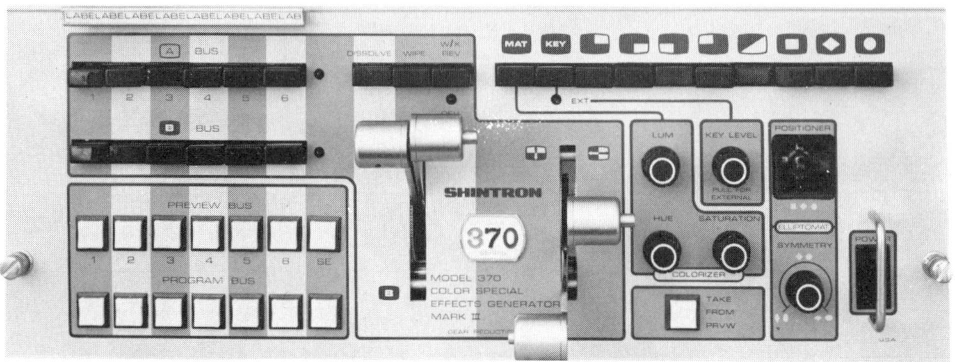

(Photograph courtesy Shintron Company)

Production switcher and special-effects generator

The switcher/fader can be a relatively simple unit, but should be considered as part of a complete console that will include the synchronizing generator, the special-effects generator, and the console monitors.

A vertical-interval production switcher is advised, because it is a single unit that combines the functions of special effects, switching between cameras, and the ability to "preview" camera shots and slides while another shot is being recorded. Vertical-interval switching also provides cleaner transitions between shots by delaying the switch to a new video source until the beam has completed a scan and has returned from the bottom of the screen to the top. Thus, the new picture does not commence in the middle of the screen, and eliminates a "flash" that might otherwise occur.

No matter what type of switcher is purchased, it should be capable of accepting more program sources than you have installed, in case you decide to bring in additional cameras for a special program.

SPECIAL-EFFECTS GENERATOR

A special-effects generator can be purchased separately, or a limited version of a special-effects generator can be included in the vertical-interval production switcher. A special-effects generator permits two different video sources to be displayed on a television monitor. The scenes can be split either horizontally or vertically; or one scene can be displayed in a corner of another picture; or insertion of shots (called "wipes") can be brought in with unique patterns.

SYNCHRONIZING GENERATOR

The synchronizing generator is the heart of the system and provides each of the camera sources with an identical set of pulses. This enables camera cuts and dissolves to be made without any breakup in the picture.

Since each camera is "scanning" a field, the cameras need to be synchronized so that when they are switched there is not a distributing "roll" in the picture caused by nonmatching signals. Dissolves from one camera to another cannot be made unless both cameras are synced to the same synchronization generator.

After being synchronized, the output of the signal from the cameras can be viewed on a waveform monitor along with the signals from the rest of the equipment, and the total system can be balanced for a stable picture.

VIDEO WAVEFORM MONITOR

A video waveform monitor provides an oscilloscope presentation of black-and-white or color-signal information. The face of the waveform monitor displays a scale of 0% to 100% to show the strength of the voltage of any part of a signal at any point. The display of variations in strength of signals allows a television-system technician to monitor the video-signal conditions to determine when adjustments are required on origination, recording, or transmitting equipment. Since all television systems desire to produce tapes with minimum distortion, a video waveform monitor should be used.

The dark shades on a television screen (called the pedestal) must be reflected in the waveform by hitting a prescribed line on the bottom half of the oscilloscope. The lightest shades similarly must be shown hitting certain lines on the top half of the oscilloscope. Since waveform monitors are part of a complete system, each waveform will have a different set of lines it must reach to determine if that system is generating proper voltage. Thus, each system has its own waveform configuration, and must be visually observed when it is known that the system is in proper working order.

The waveform monitor shows the various steps of shades of color or black-and-white (usually nine shades), and lets you know if the shades are evenly spaced. It is through the use of the waveform monitor that the technician can balance his cameras so they are generating pictures of similar brightness and contrast.

If a video waveform monitor is not used, it can appear, from observation of their pictures on a television monitor and by

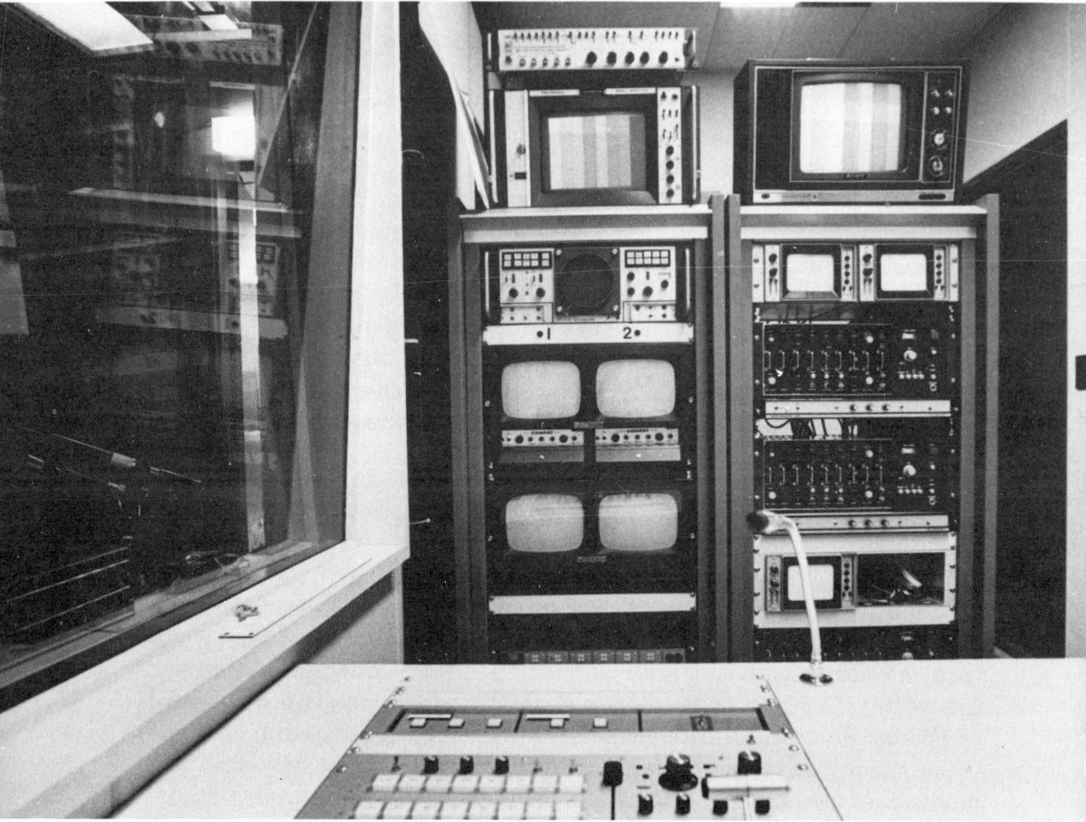

(Photograph courtesy New England Video Services)

Control racks displaying color monitors, wave form monitors, vector scope, and camera control units

switching from one camera to another, that the cameras are balanced. However, this merely means that the cameras are producing signals that project all right for that particular screen. A video waveform monitor assures you that the cameras are truly in balance and that the resultant picture will be best for the majority of screens.

TIME BASE CORRECTOR (TBC)

Videotape is an elastic medium (usually a plastic). This nonrigidity of the tape, coupled with the mechanical irregularities (bearing drag, tape friction, temperature variations, etc.) of the transport, make the uncorrected output of a VTR very unstable as compared to the

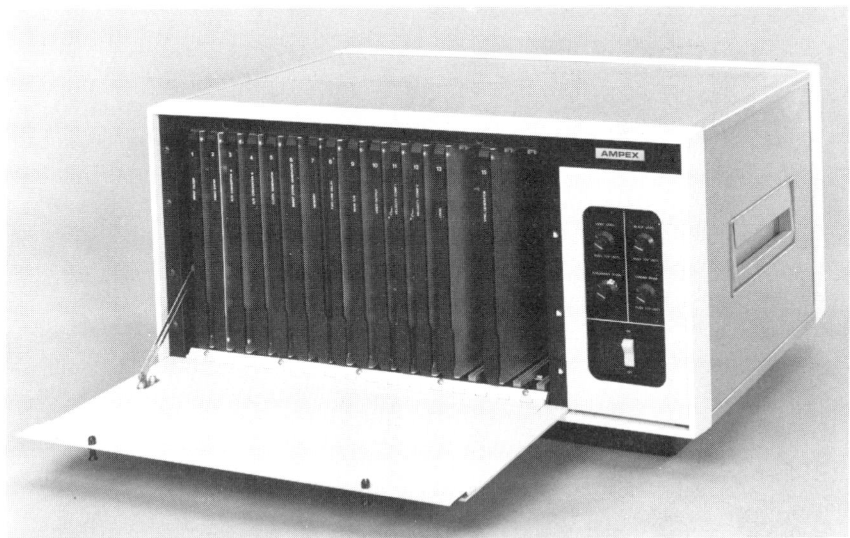

(Photograph courtesy Ampex Corporation)

Digital time base corrector. The unit produces a color or monochrome video playback signal that meets FCC specifications for broadcasting.

original input signal. As the signals are relayed (through editing, duplicating, and playback) the errors are carried forward and, usually, become more noticeable. Correction is especially essential when a producer is attempting to transfer video signals from a narrow format (like 1/2-inch) to a wider format (1-inch).

The correction of the time base is accomplished by electronic devices. These may be either analog (signal processing) devices or digital (signal conversion) devices. At the moment, digital TBCs are able to handle greater signal variations (errors).

It has been our experience that every corrective step is a trade-off for some comparable loss. In the case of the current time base correctors, the better, more stable signal is offset by an increase in picture "noise." However, the noise level is certainly more tolerable than unstable pictures.

TV MONITORS

There are two types of television sets: a monitor and a receiver. A monitor gets its picture by cable directly from where the signal originates (either a camera or VTR). A receiver gets its picture by means of a broadcast antenna, and is the type of set we have in our

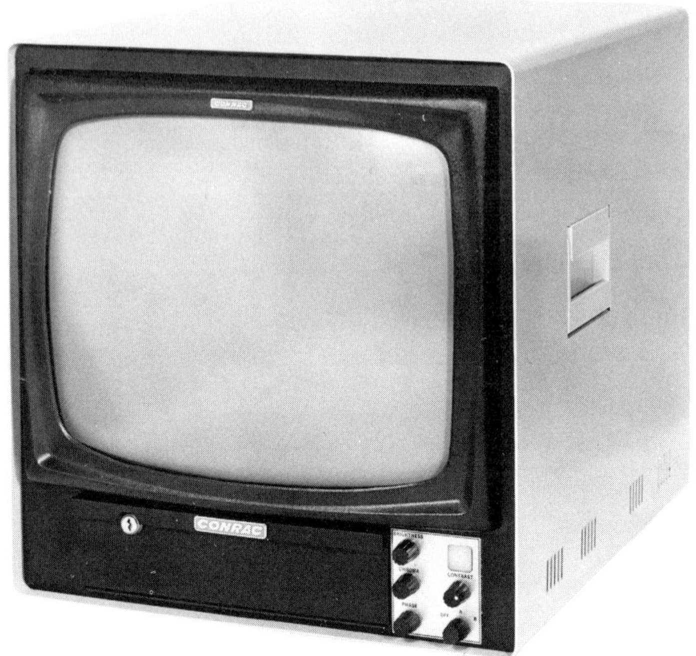

(Photograph courtesy Conrac Corporation)

Transistorized, color 10-inch television monitor in cabinet version

homes. For small-studio production, we are only discussing television monitors, although some monitors are also capable of receiving standard television broadcast signals.

In the studio, each camera will have its own monitor (viewfinder), and, in addition, a large (19- or 20-inch) line monitor should be placed so the lecturer can see and refer to visual aids picked up from other camera sources.

The control room should have one monitor for each camera; and a monitor to preview the upcoming shot; and a program (or line) monitor to display what is being transmitted or recorded. Monitors built into consoles can range from five to nine inches. The program monitor should be of sufficient size (19 or 23 inches) and of sufficient sensitivity to reflect any problems of composition or transmission that may occur.

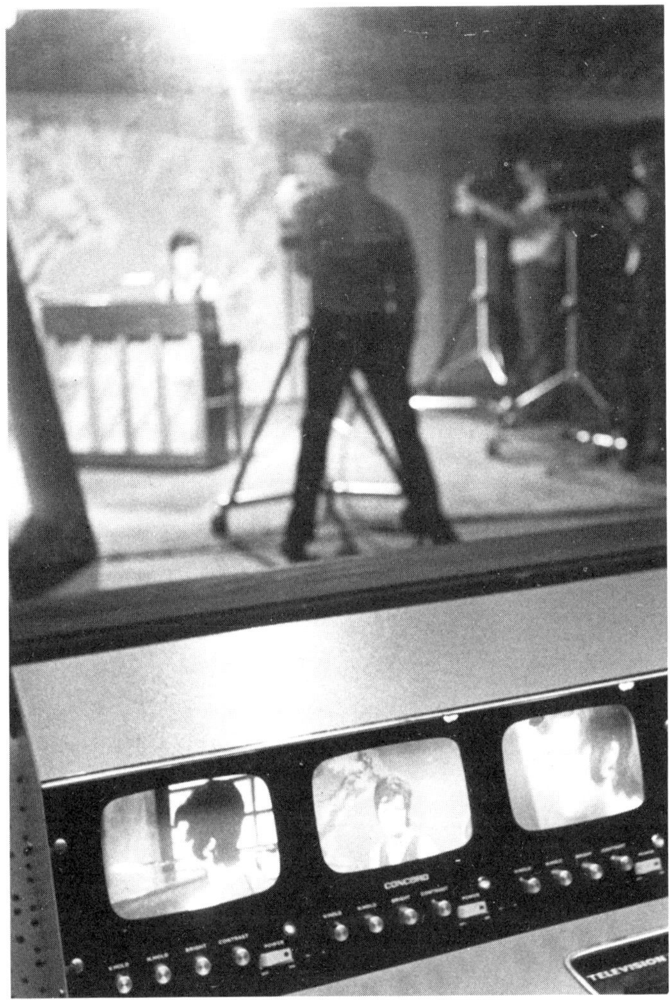

(Photograph courtesy Concord Electronics Corporation)

Camera monitors mounted in control console

AUDIO EQUIPMENT

A basic audio system consists of microphones, an amplifier, a preamplifier and mixer, and tape and record input units. Like the video system, the various audio units must be compatible, but in this case, it is a matter of impedance. Impedance is a value that is expressed in ohms. There are high-impedance and low-impedance systems that include microphones and speakers. They must be matched to avoid distortion in recording and playback.

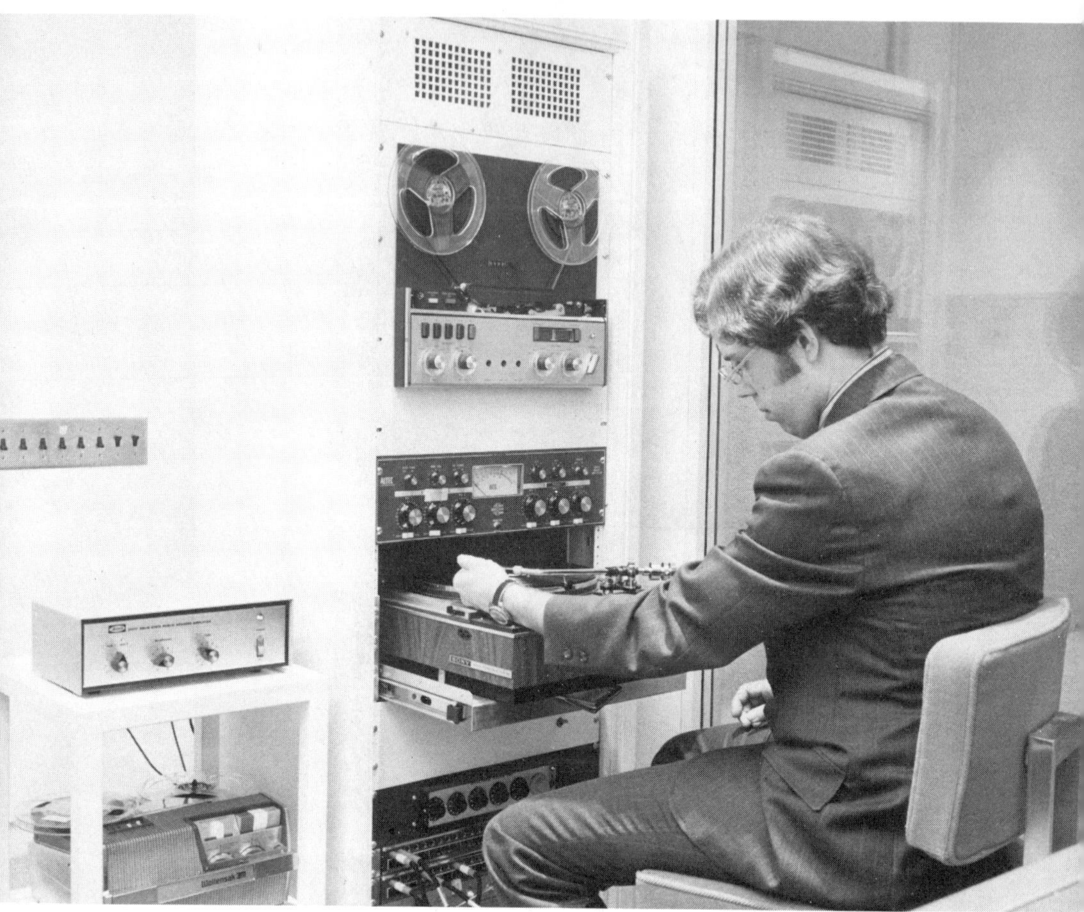

(Photograph courtesy American Republic Insurance Co.)

Audio rack in control room

Microphones are basically described by how they are supported: boom, stand, hand, desk, lapel (lavaliere). And they are further described by their range of sensitivity: unidirectional or omnidirectional. Another description is the principle upon which they operate: ribbon, dynamic, or condenser.

From our experience, we recommend the purchase of unidirectional, dynamic microphones that can be placed on floor or desk stands. Two or three lavaliere microphones should also be available if performers will be moving about during the show.

The use of desk or stand microphones eliminates the problem of a boom operator having to follow where voices are coming from, and the additional problem of occasionally catching the overhead microphone or its shadow in the picture.

The problem of desk microphones is the tendency for performers to drum their fingers on the table, or to kick the table legs, or to move the mike — all of which may be severe enough to necessitate stopping the show midway during the recording. However, a strong caution at the beginning of a session is usually enough to make a performer avoid such tendencies.

Lavaliere microphones, or neck microphones, are either hung around the speaker's neck or clipped to a lapel or tie or dress. They permit the performer to move about and leave hands free for demonstration. The disadvantage of lavaliere microphones is their tendency to pick up the sound of clothing hitting the microphone or its cable. This can be minimized if the lavaliere is not "buried" in the clothing, but is hung on the outside.

Too much emphasis is placed on trying to hide the fact that microphones are being used. A few general statements can be made about microphones:

1. The audience is aware you have to use them.
2. The audience is not curious about what microphones look like, and having established their presence on the screen, will ignore them.
3. Microphones should merely be positioned and placed at a height where they won't be in the way of movement or interfere with camera angles but will still pick up a person's voice.
4. Quality microphones are a good investment for a small-studio operation.

Audio tape recorders operate at various speeds, expressed in inches-per-second (ips), and include the following speeds: 15/16, 1-7/8, 3-3/4, 7-1/2, and 15 ips. When quality of sound is important (as your video production should require) then it is desirable to record at the two highest speeds, either 7-1/2 or 15 ips. In this way, the widest possible range of frequencies will be recorded. This is particularly important when recording music.

Included in the audio system should be an intercommunication system between the person in the control booth and the personnel in the studio. This consists of headphone communication between the production staff (which is inaudible to the talent), and a squawk-box system for the director to talk to performers during rehearsals.

All audio cables should be shielded to prevent any interference.

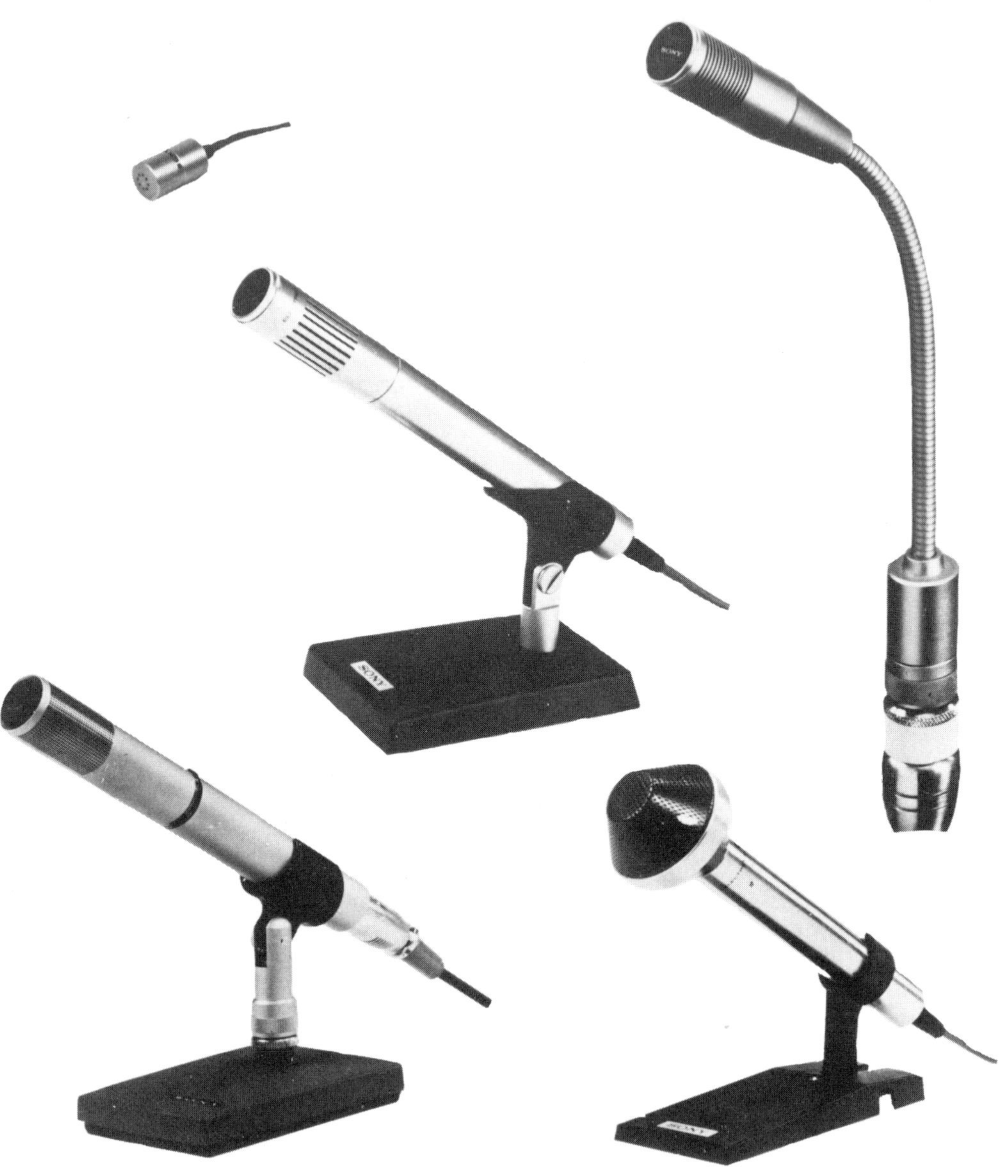

(Photograph courtesy Sony Corporation)

Microphones are available in varied sizes and shapes. The most popular type for discussions or lecture programs is the small lavaliere microphone pictured at the top left.

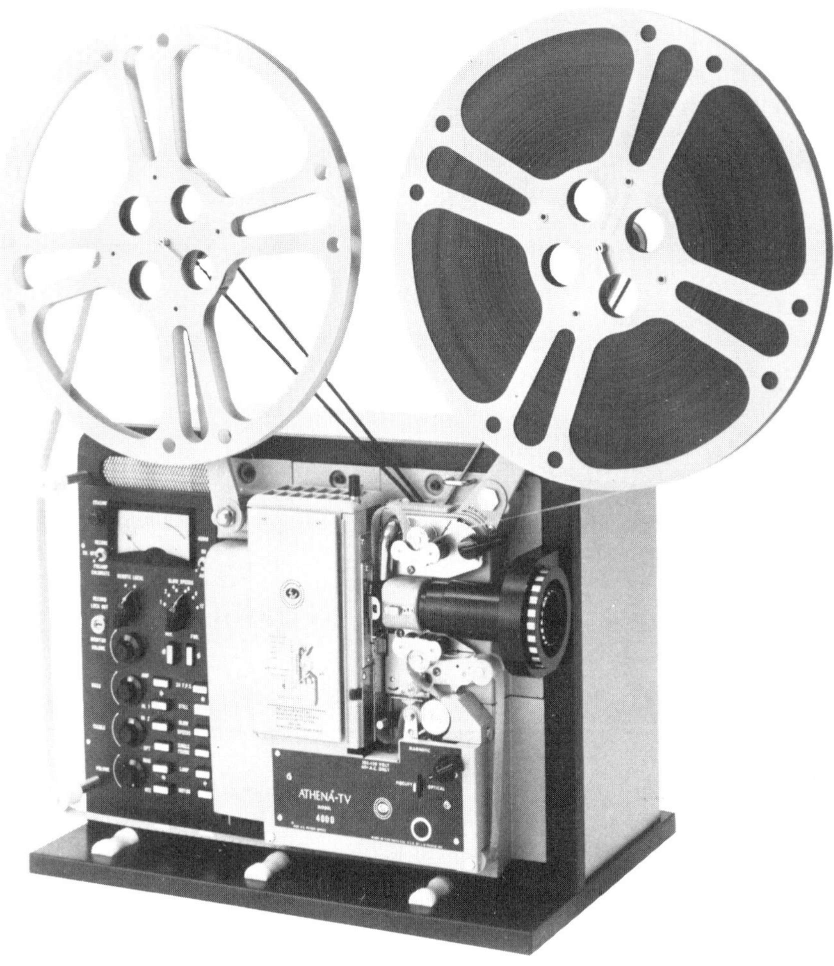

(Photograph courtesy L-W International)

A 16 mm TV film chain, animation and special effects projector

FILM AND SLIDE CHAIN

The integration of motion pictures and slides into an educational or business show is one of the best ways of giving a professional stance to the productions. This can be achieved by projecting directly in the studio (preferably using rear projection) and picking up the images

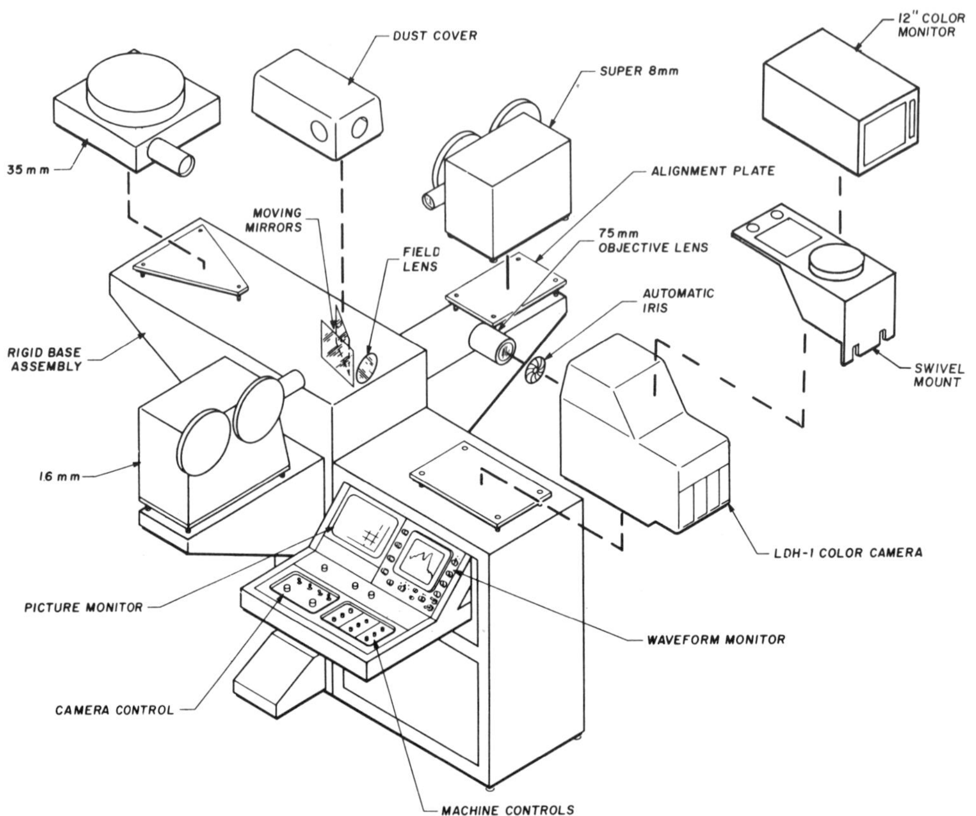

(Courtesy Phillips Broadcast Equipment Corporation)

Schematic of self-contained telecine island

on a studio camera; or by inserting film and slide transfers during postproduction editing; or by using a film and slide (telecine) chain.

A telecine chain is composed of a fixed-position camera aimed at a prism unit called a multiplexer. Standard motion picture or slide projectors are focused on the mirrors of the multiplexer that then angles the images to the video camera. The telecine camera is fed to the console in the booth and its signal mixed with other cameras. In this manner, a director can easily switch from "live" to film (or slide), or can superimpose a slide over a studio shot (such as a man's name appearing on the screen at the same time he is shown on camera).

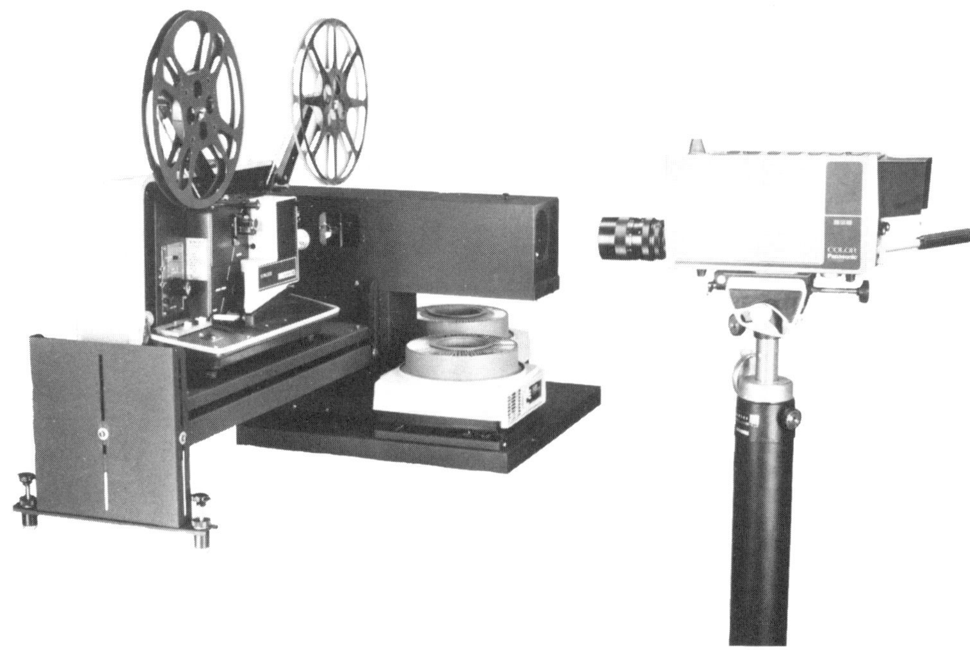

(Photograph courtesy Buhl Optical Co.)

Film/slide chain with nondedicated camera

PORTABLE SYSTEMS

Portable equipment implies those systems that are battery-powered and can be carried and operated by one person. They are generically known as "portapaks," and come in 1/4-inch, 1/2-inch, and 3/4-inch formats.

Portapak systems consist of a camera with a built-in viewfinder (that also serves as a playback monitor) and microphone; a VTR that has inputs for an earphone and external microphone (that cancels out the built-in camera microphone); a battery pack; and battery charger. They also have an AC connector that permits using the system with regular AC power. In the case of color, the system also includes a camera control unit that changes input signals to standard NTSC signals.

The most popular systems use the 1/2-inch format and weigh approximately 25 pounds. The portapaks that conform to EIA-J standards can record up to 30 minutes of programming on a single reel of tape. The batteries will last between 30 to 45 minutes, before requiring 10 hours of recharging.

A 3/4-inch portable system with camera control unit ▶

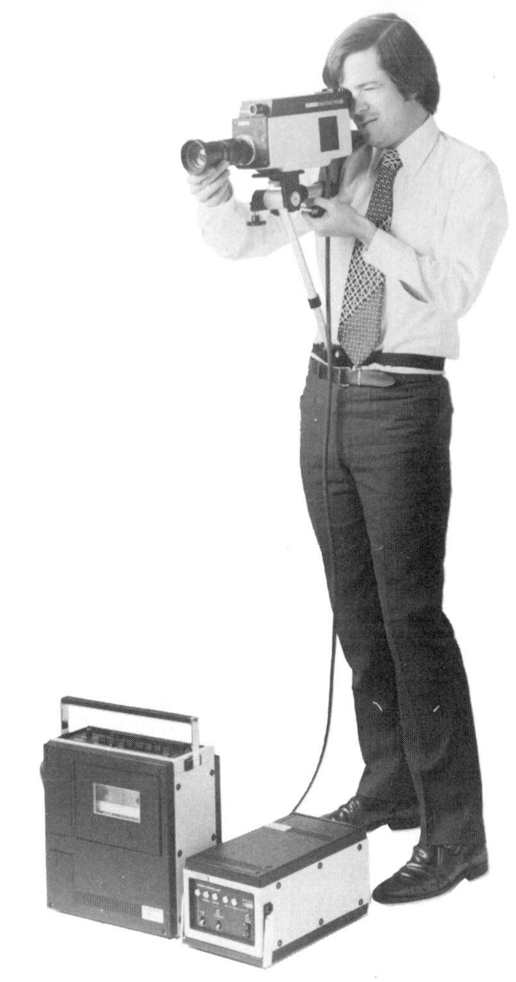

Color portable VTR system showing (from left to right) camera adapter, power adapter, video camera, camera control unit, and VTR ▼

(Photograph courtesy Sony Corporation)

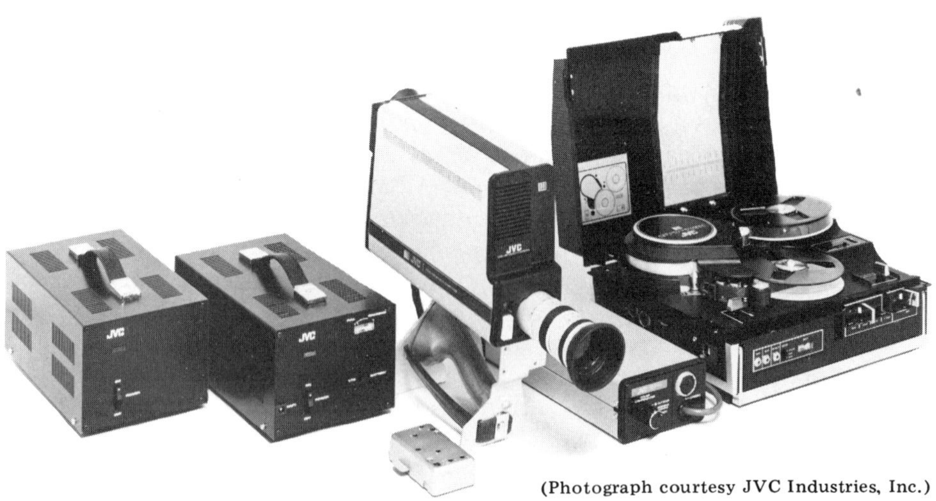

(Photograph courtesy JVC Industries, Inc.)

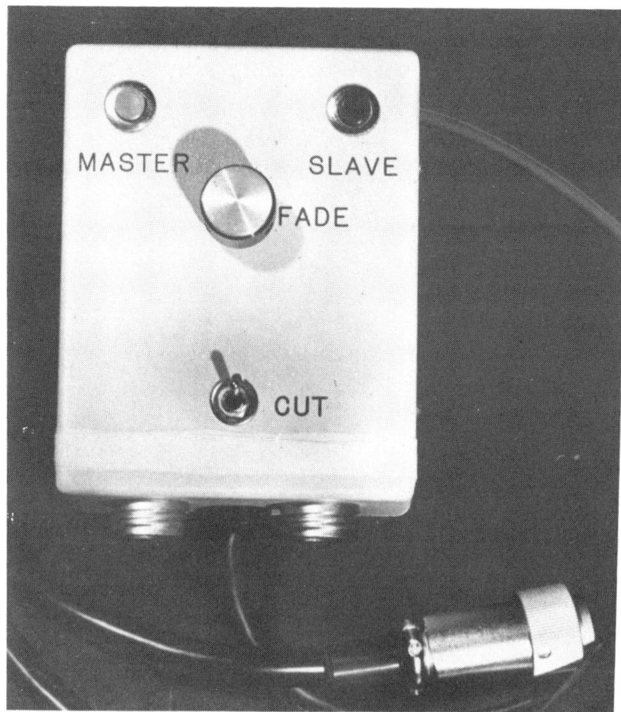

(Photograph courtesy MP Video, Inc.)

Switcher/fader used between two portapaks

A recent innovation in portable equipment is the marketing of an inexpensive switcher/fader used with two portapaks. The unit permits vertical interval switching and fading, as well as monitoring and shot-preview in the master camera viewfinder. Intercom is accomplished by using the cameras' mike jacks.

VIDEO PROJECTORS

Video projection onto large screens has been available for quite some time. Only recently, however, has the projection equipment been in the price range that warrants consideration by the small-studio producer.

There are basically two types: the fixed-size and position projector, and the variable distance projector. Picture widths range from two feet to twenty feet. It must be remembered, however, that a television picture is made up of a set number of scan lines. Thus, the larger the picture, the farther apart the lines, with the resultant effect of loss of picture sharpness.

Most video projectors are capable of projecting television pictures from off-the-air, video tape, or video cameras.

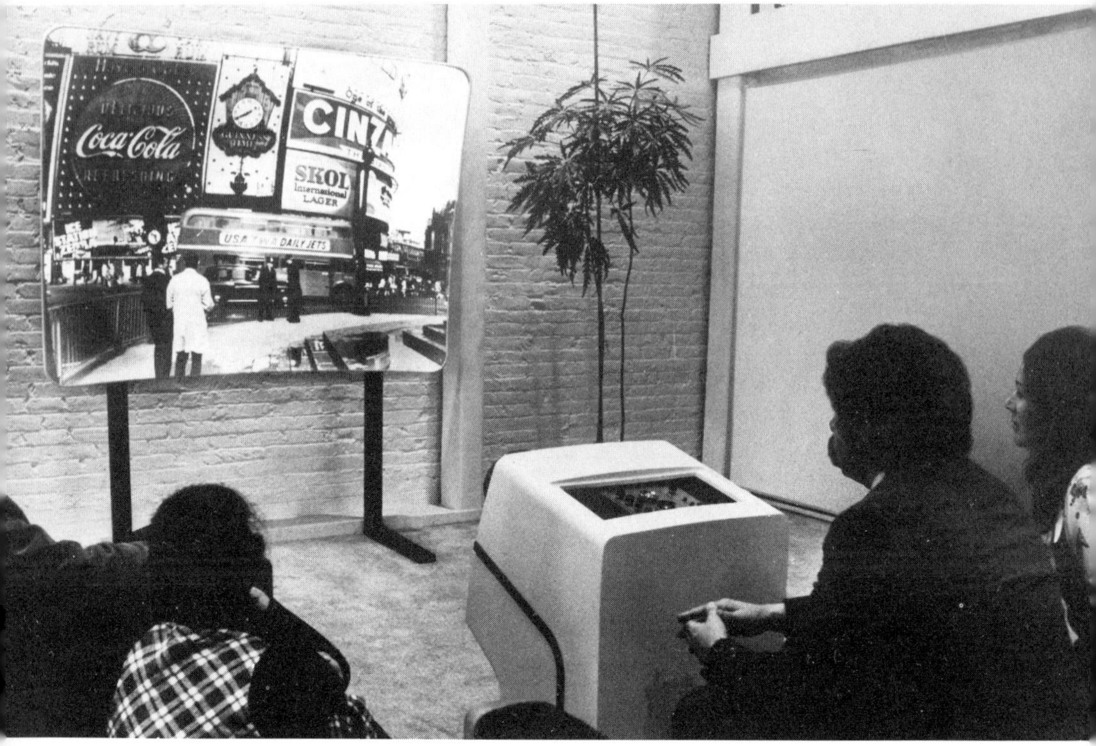

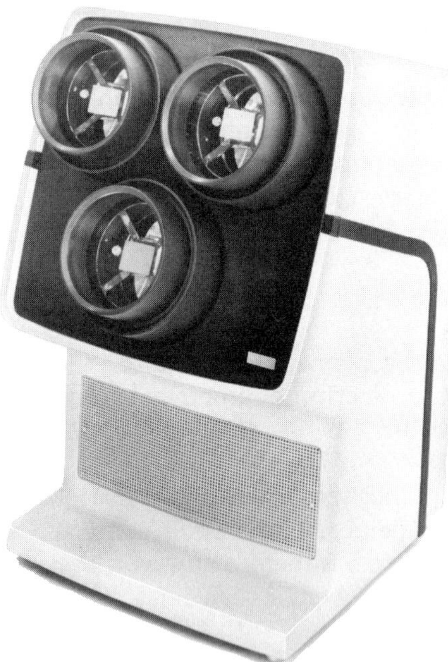

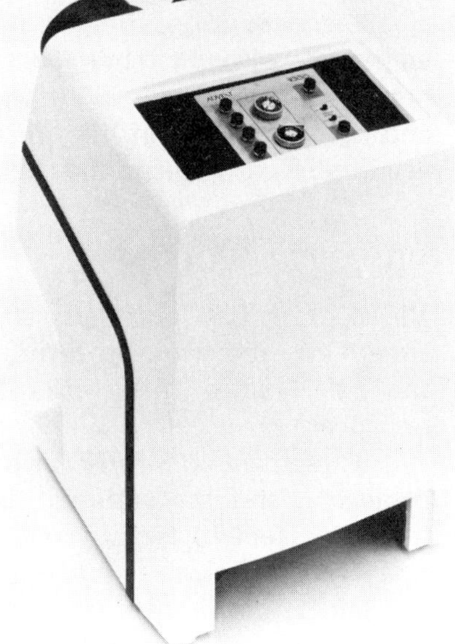

(Photograph courtesy Advent Corporation)

Fixed-position video projector utilizing a screen approximately 4 ft by 5 ft

VIDEO CASSETTE AND OTHER PLAYBACK SYSTEMS

The impact of video cassettes on the educational and industrial television markets cannot be overstated. When low-cost, nonbroadcast video tape equipment was first introduced, only the more enterprising companies and organizations ventured into production and set up their own networks. The anticipated communications "revolution" was slow, due primarily to the complicated reel-to-reel equipment involved and the lack of industry standardization.

Now with easy-to-operate and reliable playback units utilizing cassettes and the major manufacturers agreeing upon standard tape formats and playback speeds, the worldwide market is approaching billions of dollars in yearly sales of equipment and programming. Virtually tens of thousands of programs have been transferred to video cassette format and are on the shelf waiting for rental or purchase.

When first publicly demonstrated in 1969, video cassette systems were enthusiastically endorsed by educational and industrial proponents, who predicted radical changes in every phase of electronic communication. However, user acceptance was slowed by technical deficiencies, constantly revised delivery schedules, and the ultimate demise of systems that attempted to market nonstandard formats.

But now certain systems have prevailed and have established healthy, long-range markets. Other announced systems appear to have overcome technical problems and should be coming on the market within a few years.

Current systems can be divided into three basic groups. These are: (1) magnetic tape; (2) disc; and (3) Super-8mm film.

The key manufacturers include the following:

1. Magnetic tape
   - a) Videocassette (Sony)
   - b) Panasonic (Matsushita Electric)
   - c) JVC Video Cassette (JVC Industries)
   - d) VCR (Philips)
2. Disc
   - a) TeD (Teldec: AEG-Telefunken)
   - b) Philips/MCA Video Disc
   - c) RCA SelectaVision
3. Super-8mm film
   - a) Eastman Kodak VP-1

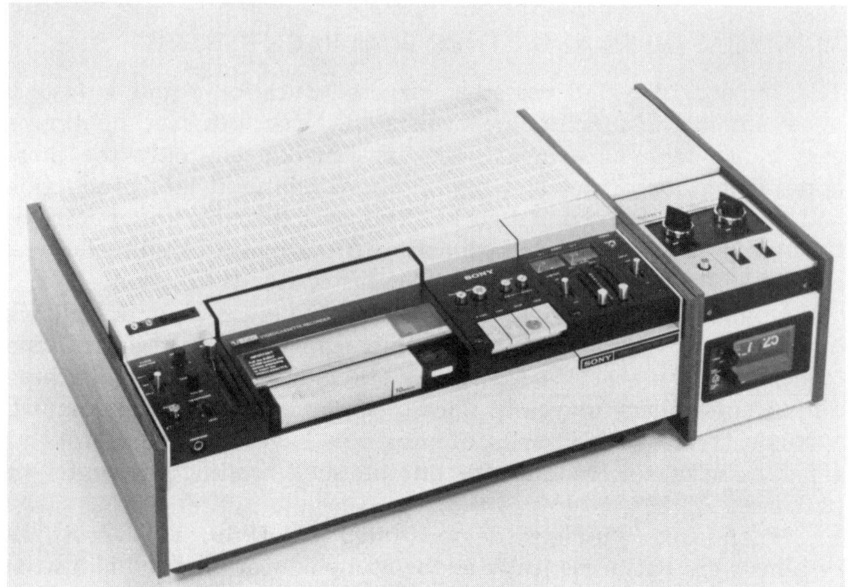

(Photograph courtesy Sony Corporation)

Sony VO-1800 3/4-inch Videocassette player/recorder

Videocassette. This system was developed and is being marketed by the Sony Corporation of Tokyo, and is based on the use of 3/4-inch (19mm) magnetic tape. The cassette itself has dimensions of 221 × 140 × 32 mm and the tape is held in two coplanar spools. The standard cassette has a playing time of 63 minutes, and the system uses 525 line/60 field television signals, with NTSC color, which is the U.S. color system.

A chromium-dioxide tape is used in this system, and moves at a speed of 9.5 cm/s. It is scanned by two video heads mounted on a rotating drum, in a conventional helical-scan configuration. Two sound tracks are provided.

The basic playback unit incorporates a modulator so that feed can be made to the antenna terminals of a conventional television receiver. The company offers an extra recording attachment that has a built-in tuner, thus enabling the direct recording of broadcast TV programs.

Panasonic. Matsushita Electric of Japan, under its brand name Panasonic, markets a 1/2-inch, EIA-J compatible cassette system, as well as a 3/4-inch magnetic tape system that is compatible with the Sony system.

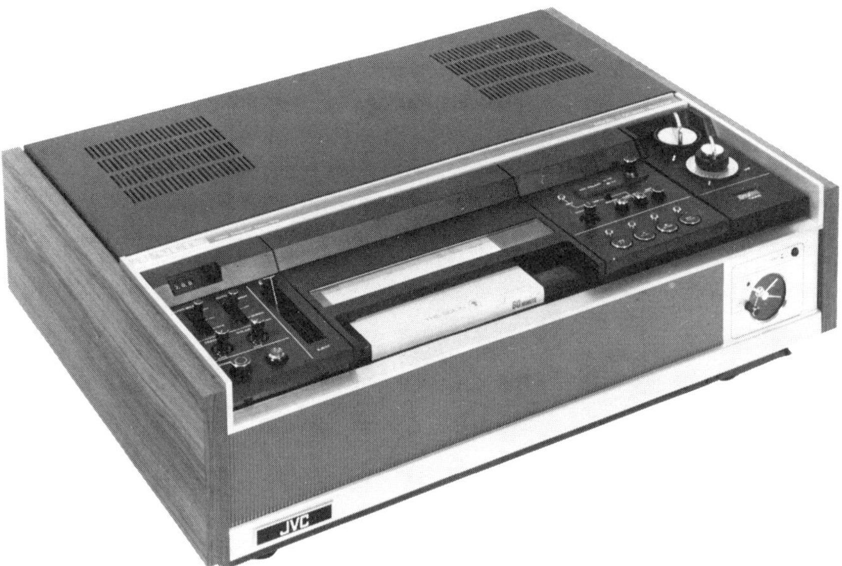

(Photograph courtesy JVC Industries, Inc.)

JVC CR6100U 3/4-inch video cassette player/recorder with built-in tuner and timer

The 1/2-inch and 3/4-inch units utilize the two-head scanning system and attach to the antenna terminals of a standard television receiver. Panasonic markets a player-only as well as player/recorder units in both tape formats. And as with similar magnetic tape systems, Panasonic is capable of recording off-the-air signals on a tape that has one-hour recording time in the 3/4-inch format, and 30-minute recording time in the 1/2-inch format.

JVC Video Cassette Systems. Another system compatible with the Sony 3/4-inch format is produced by Victor Company of Japan and distributed by JVC Industries, Inc. The JVC unit uses a two-head scanning system and attaches to the antenna terminals of a television receiver. It has a recording timer that will automatically switch on and off, permitting off-the-air recording of TV programs even when the set owner is not present. It is also equipped so that the viewer can watch one program on his set and record for later playback another program being broadcast at the same time over a different channel. An additional feature is an automatic replay button.

VCR. This video cassette system, developed by Philips, of Eindoven, Netherlands, is based on the use of 1/2-inch (12.7mm) magnetic tape. The cassette has dimensions of 126 X 146 X 41 mm and holds

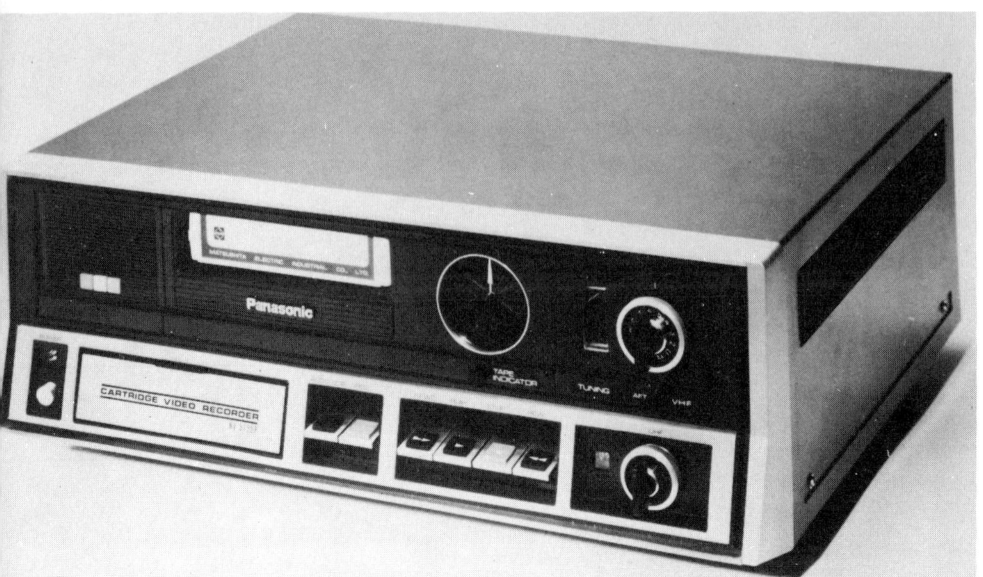

(Photograph courtesy Matsushita Electric Corporation)

The Panasonic NV-5125, 1/2-inch video cartridge player/recorder with VHF/UHF tuner

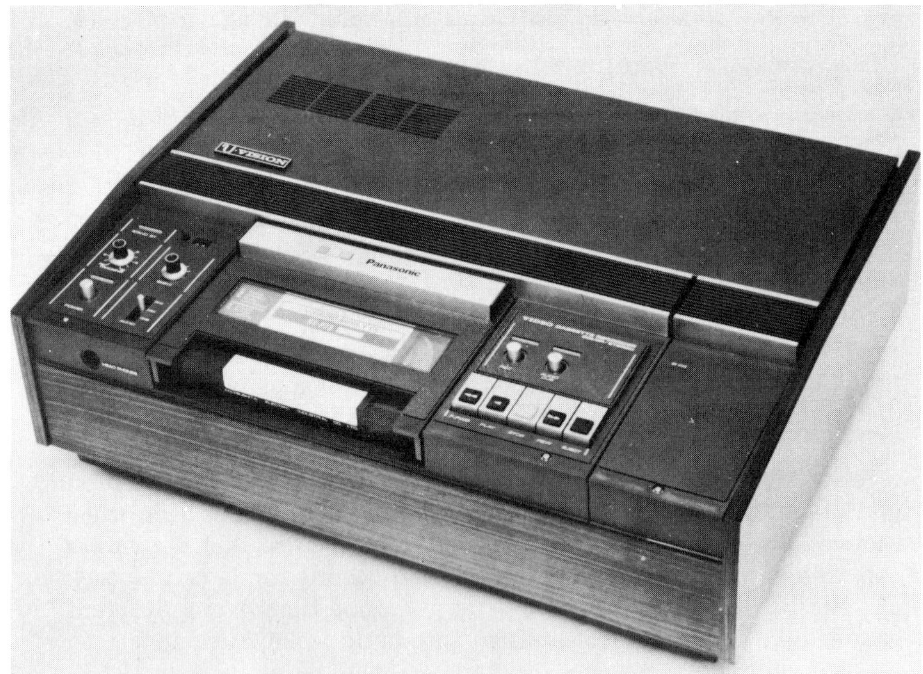

(Photograph courtesy Matsushita Electric Corporation)

The Panasonic NV-2120, 3/4-inch cartridge player/recorder

the tape on two superimposed spools. The playing time of the cassette is 60 minutes.

Chromium-dioxide tape is used in this system; it moves at a speed of 14.3 cm/s and is scanned by two video heads mounted on a rotating drum, in a conventional helical-scan configuration. Two sound tracks are provided.

At present, the system operates using 625 line/50 field video signals with PAL color, which is not compatible with the United States color system.

VCR incorporates a VHF/UHF tuner, so that regular TV programs can be recorded directly, and utilizes a modulator so that in the playback mode it can be fed directly into the antenna terminals of a conventional television receiver.

In the recording mode, there are automatic level controls for sound and picture. In playback there is a stop-motion capability for examining fast-moving sequences in detail. The system also has a built-in timer for automatically switching the equipment on and off at predetermined times.

Philips has announced that VCR has been agreed upon for standardization by a number of other companies, including AEG-Telefunken, Blaupunkt, Grundig, Loewe-Opta, Nordmende, Saba, Sony (Europe), Studer, Thorn, and Zanussi.

TeD. This video disc system is manufactured by Teldec (AEG-Telefunken) of Berlin and London. It was first demonstrated in 1970 and was marketed in Europe beginning in 1975. The present playing time of a 12-inch disc is 12 minutes. The playback equipment resembles a portable record player.

The record rotates at the rate of 1500 revolutions per minute. Information is recorded on the disc as a frequency-modulated signal and is reproduced by the operation of a diamond stylus in combination with a pressure-sensitive ceramic pick-up. The spiral groove has an extremely fine pitch, having about 130 grooves to the millimeter.

Originally developed as a monochrome playing system, color has been added to the system.

One of the advantages of this system is that the mass production of duplicates is very inexpensive. The discs are made by conventional phonograph-record-pressing techniques. The recording technique is likewise much the same as phonograph record cutting, but it is necessary to reduce the bandwidth of the input signal by extending the time scale during cutting. Because the production process is elaborate, the system is limited to prerecorded material, and it is unlikely that recordings will ever be made by the user.

Videodisc has one sound track and the average life of the recordings is expected to be on the order of 1000 playings.

Philips/MCA Video Disc. In September 1974, MCA Inc. of California and N.V. Philips of the Netherlands announced agreement for the joint development and sale of optical video disc players and compatible discs. The video disc player would be manufactured and marketed by Philips, while MCA would manufacture and market video disc programs. This announcement was significant, since prior to that date the two companies had worked on independent, noncompatible systems.

This new system will playback through a standard television receiver on a selected RF channel by a connection to the antenna terminal. The player will pick up the video information off the disc by means of a low-powered laser beam.

The advantage of the Philips/MCA system is that it will be capable of a playing capacity of 60 minutes per disc, with speculation that may be increased to 120 minutes by recording on both sides.

As of the time of publication of this book, this partnership had not established equipment delivery schedules or other licensing agreements.

RCA VideoDisc. In March 1975, RCA Corporation announced a video disc system to "be ready for production by mid or late 1976." The RCA system will record on both sides to permit 60 minutes of programming. The player will use a sapphire-tipped stylus in the pickup arm to follow a tiny spiral groove on the disc. However, the actual sensing of the signal is not done by pressure. Instead, a thin metallic electrode on the face of the sapphire senses variations in electrical capacitance created by irregular slits in the bottom of the groove.

RCA announced that the master records will be made by an electron beam focused on the groove in a sensitized disc. Bits of the sensitized materials that are touched by the beam are then removed, leaving crosswise slits of varying width separations within the groove. From this relief pattern, a mold is made for stamping out discs. The final recorded discs will then receive a thin coat of materials to enhance the variations in electrical capacitance and a film of oil to lubricate passage of the stylus.

RCA anticipates that a disc should last for at least 200 replays and that the stylus should last for 300 to 500 hours.

Eastman Kodak VP-1. The uniqueness of the Kodak Super-8mm system is the development of a plastic film cartridge that can be interchangeably used on an optical sound Super-8mm projector or on a video player. Thus, the single cartridge with the Super-8mm format can be shown in a variety of facilities depending upon the availability of projection equipment.

The video player is in effect a color or black-and-white film chain that can also be used to insert Super-8mm film clips into longer video tape productions or as a unit for film-to-tape transfer. The cartridge will take open-reel material up to 22 minutes in length, and no special processing is required. The video player has the capability of running at speeds of either 18 or 24 frames per second. It also has the capability for still-frame or single-frame advance.

PURCHASING

It is extremely advisable to purchase the equipment from a local distributor or a national manufacturer who has a locally franchised maintenance outlet. Before you place your orders, be certain that

- the equipment purchased from various manufacturers is electronically and mechanically compatible and will interface with all the other pieces of hardware;
- the components are built to the Electronic Industry Association (EIA) standards;
- the complete system, not just the individual components, is guaranteed by the supplier.

It is possible, of course, to find a lot of "bargains" in used television equipment or off-brand units. However, any cost savings are later lost while your equipment and personnel stand idle waiting days or weeks for a replacement part or qualified technician to show up.

If you purchase from a local distributor, who can supply you with a complete replacement unit while your unit goes into his shop for repairs or preventive maintenance, no time is lost. The local dealer is more inclined to give quicker service and stand behind warranties.

All VTRs should be located in areas that are dust-free. Dust on video heads can cause dropout and excessive head wear, and dust on a tape can cause tape "noise." And, of course, any strong magnetic fields close by can cause VTR components to become magnetized, and, similarly, can cause degradation of magnetic fields on the recorded tape. (A discussion of care and maintenance of equipment will be found in Chapter 17, Postproduction Effort.)

Learning to operate a video tape recorder and cameras is not unlike becoming acquainted with any new semisophisticated mechanism. The first few ventures will be relatively simple ones following the instruction manual that accompanies the equipment. Once the basic steps are mastered, you can then move on to the more courageous and innovative steps that the system is designed to perform. From that point on, the operator can bring to bear all the talent — and supplementary equipment — desired so as to broaden the total system's capability.

Thus it can be assumed that the first few programs to be produced may not have public showings — they may be strictly shakedown cruises for the production staff. And even if produced with the intent of audience acceptance, the first few shows may be too amateurish for release. The point, of course, is that the purchase of the equipment does not necessarily mean you are ready to enter the video production business.

CHAPTER 5

PRODUCTION PERSONNEL

How many persons does it take to produce a continuing number of television shows? In a nonnetwork, commercial telecast, there are a minimum of 12, not including the talent. In a small-studio, nonunion production, the number, of course, varies with the variety and frequency of productions. Each show that is distinctly different from another show must be planned, scripted, designed, rehearsed, and shot. However, once a format is established, some of the steps can be eliminated and most steps can be shortened.

A periodic, ad-lib discussion program can probably be shot in two or three hours, involving a full-time coordinator/director, a technician, camera personnel, and the talent. A fully scripted training program utilizing a number of visual aids can take two weeks to prepare, one full day to shoot, and involve over 10 full-time and part-time people.

Certain assignments, however, are essential to the production of a television show, no matter what the size of the studio and the frequency of production. Each assignment must be accounted for, whether it is an individual's full-time job (such as a video technician), or it is on an "as required" basis (such as a graphic artist).

At the end of the chapter are all the assignments normally associated with a commercial network video tape production. Although a small studio may need only a fraction of the personnel a commercial production requires, the assignments must nevertheless be fulfilled even if handled only by one person.

Certain assignments can be combined, such as a producer/director or writer/producer. This, of course, can be overdone. It would have to be a truly exceptional person who could effectively handle the producer/director/writer assignments. It would be wiser to find a good writer and coach him in television production technique, and leave the creative interpretation and implementation in the hands of someone else.

Producer. One of your first considerations before entering the television field is who will be responsible for the production of your video tapes or your closed-circuit programming. That person is the producer and is ultimately responsible for the quality, content, and distribution of the show.

Unlike commercial television, the producer of an industrial or educational television show will probably maintain a staff position within the organization. The amount of television production may warrant his or her full-time attention or may allow the performance of other duties such as audio/visual director, training director, instructional supervisor. The producer could be the head of public relations or the director of marketing, depending upon the range of programming.

One very important point regarding who is designated as the producer: that person must have the authority to say "yes" to the production personnel and the authority to say "no" to other department heads. The producer must be able to respond quickly to technical problems that arise and be authorized to change locations, purchase equipment, bring in additional personnel. He or she must also be authorized to make purely subjective judgements on scripts, talent, and visual materials. The producer must have complete authority over the total production.

The producer may either originate a production idea based upon personal knowledge or develop a production requested by another individual or group. The former situation may require some selling on the producer's part; the latter situation requires the successful execution of the idea to the satisfaction of the originators and the audience.

Such elements of production as the selection of personnel; the determination of locations and settings; script approval; selection of graphic design — these and all other nonshooting elements are the responsibility of the producer. A major responsibility is establishing a production budget, and seeing that costs are kept within that budget.

Writer/Researcher. As we indicated earlier, any good writer can be taught television techniques, but not all producers and directors can be taught how to write well. The main problem in writing for television is the tendency to oververbalize — not permitting the visual aspects to carry the main part of the show or to permit an audience to participate in what it is viewing.

The requester of a show should not be permitted to write the same show. That person could not objectively select what would appeal to the audience, but would tend to select what was of personal interest. Nevertheless, a knowledgeable person must be assigned to the writer to supply the information, rather than having the writer attempt to ferret out the information alone.

A good public relations writer is usually a safe bet if a show must have a script. (If it has no script, the director can determine the integration of movement, demonstration, and visual aids.) A public relations writer is used to doing his or her own research and has the ability to interview qualified personnel on a subject and select the items of audience interest. A writer has also probably had some exposure to audio/visual presentations and is aware of the importance of interesting images and creative graphics.

Director. The key figure in the actual production cycle is the director — the person who makes the major decisions regarding the shooting of the show and who assumes the major responsibility for successful execution of programming.

That person, quite literally, calls the shots. In addition to the planning and scheduling necessary to begin a production, the director may also be required to produce programming, which means assuming total production responsibility.

In a small-studio situation, the director will be the most important part of the production process and may also have administrative (plus creative) responsibilities. It will be the director's responsibility, for example, to advise the producer on the scheduling of facilities so as to yield maximum use.

Scheduling will, of course, include actual production time as well as rehearsal time. While it is quite possible to rehearse material in areas other than the studio, a final rehearsal must take place in the studio. It will be at this juncture — before the final rehearsal — that the director will be able to prepare shooting notes to establish the continuity of the story to be told and the precise placement of the individual components of the program.

An illustration is probably in order.

When the program's participants are ready for the final rehearsal in the studio, the director will proceed through the action of the proposed program with the individuals and all of the intended props, sets, and visuals. The cameramen will be present so that they can likewise be alerted to any special requirements or problems.

As the intended action of the program is rehearsed in detail, a log of camera shots (a "shot sheet") can be made by the director for later use in the control booth. This log will include pertinent action, the points at which the performers will make a significant movement (such as standing up, walking, etc.) or gesture (pointing at a model, a drawing, a blackboard, etc.). It will also contain indications where video tape inserts are to be used, film clips or 35mm slides (if capabilities exist for their projection), prerecorded audio information, and studio visuals such as charts, photographs, etc.

It is important that this step be worked out carefully since it is easy to get cameras in awkward positions, not only in terms of camera usage and movement (crossing camera cables, and getting one camera in back of another one), but also forcing a camera to use an inappropriate lens, thereby creating distortion or some other unwanted effect.

The more carefully planned and rehearsed the sequence, particularly in a training program, the more professional it will appear in finished form.

During the actual production, it is the director's judgment that determines the choice of shot and the relationship of the pictures and sounds that are being recorded on the VTR. More of the director's role is examined in Chapter 8 relating to direction. At any rate, it is the director who has complete charge over the activities of the technical staff working on a given program, directing and coordinating the activities of the camera operators, and any of the technical personnel who may be working in the booth (film chain, riding gain on audio, video controls, etc.).

Technical Director. A technical director who works a show will work in conjunction with, and take orders from, the director. He or she may also be called a "switcher"—one who physically does the switching (dissolving, cutting) from one camera to another. If the technical director does this, the director is then free to follow the script, the shooting log, or any notes that may previously have been made regarding the program and its content or organization.

In situations in which the switcher is used, that person may also operate the controls of the microphones (riding their gain or volume) and also may make minor contrast or color adjustments (or advise the camera operators to make minor adjustments) on the individual cameras themselves. A switcher will not be able to operate a film chain particularly if there are films to be threaded or projectors to be loaded during taping. The technical director, if used, is glued to the position in front of the switching equipment.

Cameraman. The camera operator plays a key role in video tape production. It is the operator's care and attention to detail that are most important—also skill, in the basic handling of the equipment.

Camera operation is not a difficult skill to master. It is simple. It is almost as though an instantaneous movie were being shot and projected on a private screen. It is this very capability that makes it easy for the novice cameraman to "learn" visual skills — learn to look *with* the camera, learn to follow focus, learn to smoothly operate the controls of a zoom lens.

As with any other discipline, the cameraman learns about camera operation *by doing*. As we mentioned, it is a painless way to learn about photography. There are no technical mistakes that can be made (such as the choice of film), and the result is immediately available in the viewfinder. If there is insufficient light on the subject it can be remedied. If a subject is out of focus, the operator can correct the problem.

From our own experience we can say that a cameraman must log a certain amount of time with the camera — learning its limitations, learning to use the medium. The operation of a particular camera should become second nature, and its proper use, strictly a matter of habit.

It is essential that any would-be director be as proficient as possible with a camera, so that he or she can note the possibilities (or as importantly, the limitations) of the available equipment, materials, lighting, and studio space. More about camera technique is explained in Chapter 9, but one of the additional duties of the cameraman (if there is no floor manager) will be to signal the talent when the camera is about to be activated.

One may do this by extending an arm overhead just prior to the point when the director "takes" that camera, at which time the arm goes down and the red light on the front of the camera goes on. This gesture gives the participants ample forewarning that the camera is about to be changed so that they can comfortably begin "playing" to the other camera . . . that is, focusing their attention, or their activity to the taking camera.

Engineer or Technician The question of who will *operate* the expensive video equipment is often one that is overlooked or minimized. Many manufacturers suggest that virtually anybody can run the equipment, and this is technically correct — almost anybody, for example, can point and focus a television camera, but it requires considerable training to be able to maintain the camera in constant working order.

Effective and efficient production requires the application of creative people who are concerned with program content, continuity, and pace in television productions. But effective production also requires that somebody be around who knows how to do the necessary engineering and perform the needed technical work.

The care and maintenance of equipment is absolutely vital to protect the capital investment.

At some point, it must be decided who will be responsible for the electronic quality of the product. It may be the director or somebody else who is given special training (often by the manufacturer) in the care and treatment of the camera and recording equipment (plus the other electronics in the system).

The alternatives are these: (1) a resident technician or engineer, (2) a member of the creative staff who is trained for this work, or (3) a local or regional service organization that can visit on a prearranged schedule to do the necessary trouble-shooting and maintenance.

During the actual taping, the technician can assist the director in handling certain equipment such as the film chain or audio mixer. In postproduction, a technician may also serve as tape editor.

Designer and Artist. As you will see in the chapters devoted to television graphics and set design, the problems of television are different from those in any other medium. The successful designer or artist for this medium is going to have to gain experience in finding combinations of colors or values that work best in given equipment and lighting situations. He or she must work closely with the production group and understand their unique needs. And like the cameraman, the designer must work closely with the medium (in this case by simply viewing a monitor) until he or she discovers which designs and colors work best in a particular system.

Floor Manager. A floor manager can be quite useful in two major ways. In preproduction, he or she serves as liaison between the director and the participants and is in charge of the many details of dressing the set, making sure that visuals are on hand in the proper location.

During actual program shooting, the floor manager is in the studio and wears the same type of headphones worn by the camera operator, to make it possible to listen to the director's instructions and know, ahead of time, what camera will be called. Then, from appropriate gestures (pointing from one camera to the next slightly before the switch is made), the performers are alerted to the imminent switch from one camera to another, and can therefore prepare to focus their attention and action toward it.

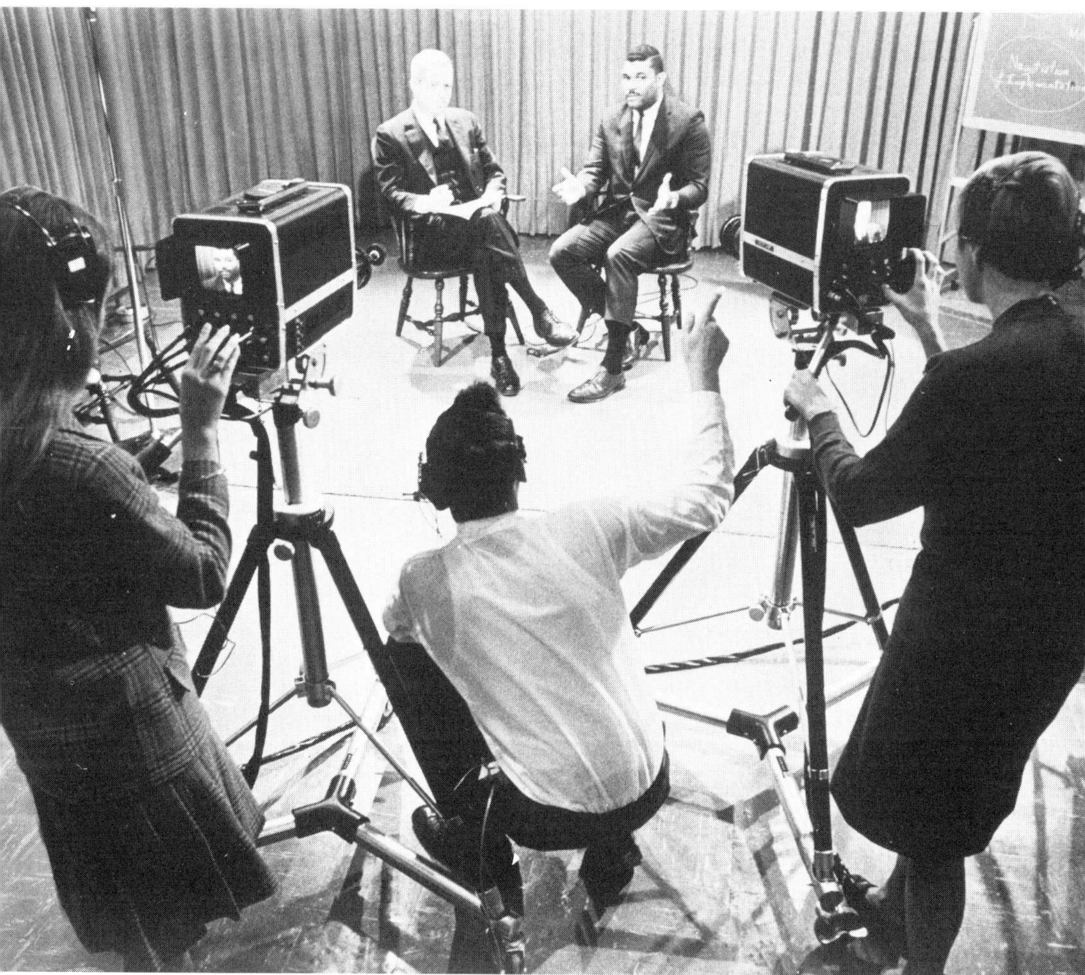

(Photograph courtesy Matsushita Electric Corporation of America)

Floor manager prepares to cue talent on set

The floor manager can also place properties on the set, and can handle flip-charts as they are being shot. He or she also may have certain postshooting activities to perform, such as striking sets, storing props and visuals, and cleaning the studio.

At one point, it was considered standard in a regular broadcast studio to begin one's production employment as a floor manager, and advance to other, more demanding jobs (assistant cameraman, mike boom operator, cameraman, and perhaps, eventually, director). The apprenticeship was useful in providing experience in the practical matters of television production and in achieving certain types of programming results. Such an apprenticeship can also apply to a small-studio production facility.

NETWORK COMMERCIAL PRODUCTION PERSONNEL

Executive producer — responsible for the creation and production of a series of shows.

Producer — responsible for the creation and production of a single show.

Associate producer — assistant to the producer or director in handling details other than the actual taping.

Production assistant — handles such details as clearing music rights, delivery of graphics, etc.

Unit manager — acts as business manager in seeing that costs are kept within the production budget.

Writer — develops the script from the basic concept.

Research assistant — researches material and checks accuracy of information.

Director — determines the creative aspects of the show and directs the cameras during the taping.

Assistant director — responsible for the timing of the show, cues in special effects and music, and prepares the camera shots.

Floor or stage manager — directs the activities on the floor of the studio and cues performers and stage hands.

Technical director — In charge of all technical personnel and runs the console during the taping.

Audio director — responsible for all audio aspects of the production including microphones and audio tapes.

Records person — handles all audio disc problems including music and sound effects.

Video director — handles master video tape recorder including shading of pictures when recording in color.

Cameramen — set up cameras under direction of technical director and run cameras during taping under direction of the director.

Boom operators — responsible for boom microphone in following talent during taping of show.

Audio assistant — moves dolly into position to assist boom operator.

Chief grip — supervises stage crew during setting up of production and actual taping.

Crane operator — controls dolly if crane TV camera is used.

Film chain operator — responsible for equipment that integrates motion picture footage and slides into production.

Tape editor — responsible for postproduction editing of tape.

Lighting director — determines basic lighting for the show and supervises setting of the lights during technical rehearsals.

Scenic designer — creates basic design for the setting and supervises construction.

Set decorator — carries through on completing stage design with props, etc.

Music director — determines music to be used and arranges for recordings or musicians.

Graphics director — creates and executes graphic designs such as slides, titles, etc.

Makeup — applies makeup to talent.

Talent — on-camera personnel, announcers, narrators, etc.

No successful video production was ever the product of mere chance. Smooth, professional programs are the result of planning, rehearsal, and attention to detail during the recording session. This implies the selection of people with the proper training and skill, and the freedom to apply their knowledge to get the job done.

CHAPTER 6

PRODUCTION PLANNING

The intent of this chapter is to discuss all of the various aspects that must be considered before a production can be undertaken. It tries to solve problems (ahead of the actual production process) by simply asking enough questions. Is there enough information to *do* a show? Will the right people be available? Do we have the right equipment? Are there enough electrical outlets? Can we get a camera close enough to the needed lab set-up? Can we develop a reasonable budget for the production? Can we stay within the budget once we develop it?

This, in other words, is a nuts-and-bolts chapter that discusses all of the background effort that must be considered and planned before the "creative" work can ever be done. The how-to part of production is discussed in the chapters on direction and the production process.

Television production is a complicated business, and contains a lot of activities that may seem trivial but which, when taken together, become vital. Successful administration is the key to it.

There are two important jobs that must be done in order to administer a small-studio production operation. The first is the general practice of keeping records of the work actually accomplished (as well as that which is scheduled or planned).

The other critical job is to be able to estimate adequately the costs of doing business. *Doing business* is not a bad way of viewing the problem. The small-studio operation can very easily be viewed as an independent business, with all of the costs inherent to operating. There are costs of building occupancy to be considered and depreciation schedules to be drawn up for capital equipment.

To assist in the planning job we have developed a form which we call a "Television Production Planning Sheet." You can easily modify this sheet to your specific needs, eliminating some of those elements you're certain you will never need, and adding certain specialized things of your own — elements that will continually reappear in your plans or estimates.

The sheet serves the dual purpose of providing a history of the job: i.e., the title of the program, the nature of the program (including content), the date produced, the key people connected with it (both in front of, and behind, the camera), and the costs incurred (or the time spent). It also tells who approved the production in the first place, and shows the account number against which the production was done.

One of the techniques we have employed has been to have the basic form information printed on the front of a manila folder (with the words "production number," "title," and "date" also printed on the tab portion of the folder). In this way it is possible to keep all records of the production, including scripts, bills, time charges, and other data in one place. This is particularly important if time and materials must be shared among various projects or accounts. It is a useful technique when it comes to computing bills for production expense. It also serves as a reminder that *all* accrued costs are recorded against the production. It is very easy to *lose* accountability by failing to charge for legitimate expenses. One such expense that often slips away is the cost, sometimes quite considerable, of typing, clerical help, and reproduction.

Remember that if one is to be accountable financially for a production operation, he must keep track of every dollar in the budget. There is little sense in "fudging" numbers or burying costs, because you eventually get a misconception about what things *really* cost, both in time and materials.

Item 1 is obvious; it is the date on which the planning sheet is executed.

2. Production #. The production number can be handled in a number of ways, but the way we have chosen is an alpha-numeric system in which a letter of the alphabet is followed by a number. The letter is related to the requester of the material. It could be the initial of the last name of the individual, the initial of the requesting department (i.e., manufacturing, research, training, biology, etc.), or it could stand for the initial of the client organization requesting the work.

The numbers, then, follow in serial order and indicate the number of productions done for that individual, department, division, or company.

Television Production Planning Sheet	(1) DATE			(2) PRODUCTION #
	(3) TITLE			
	(4) REQUESTED BY			
(5) DESCRIPTION OF SUBJECT				
(6) ESTIMATED RUNNING TIME	(7) INDOCTRINATION REQUIRED YES___ NO___			
	(8) TIME	DOLLARS		DATE
(9) PRODUCTION				
(9.1) IN STUDIO				
(9.2) ON LOCATION (IN-HOUSE, OFF-SITE)				
(9.3) OUTSIDE FACILITIES				
(10) EQUIPMENT/POWER REQUIREMENTS				
(10.1) EQUIPMENT IN STUDIO				
(10.2) PORTABLE EQUIPMENT				
(10.3) RENTAL EQUIPMENT				
(10.4) ELECTRICAL POWER NEEDS				
(11) PREPRODUCTION CONSIDERATIONS				
(11.1) PRODUCTION PLANNING				
(11.2) SCRIPT OR OUTLINE PREPARATION				
(11.3) TYPING, REPRODUCTION, MESSENGERS				
(11.4) PRODUCTION OF VISUALS (TIME/MATERIALS)				
(11.5) PHOTOGRAPHY				
(11.6) REHEARSAL OR WALK-THROUGH				
(11.7) AVAILABILITY OF TAPE, ETC.				
(11.8) SETS				
(11.9) PROPS				
(11.10) COSTUMES				
(11.11) MAKEUP				
TOTAL, THIS PAGE:				

SHEET 2	TIME	DOLLARS		DATE
(12) POSTPRODUCTION REQUIREMENTS				
(12.1) EDITING				
(12.2) NUMBER OF DUBS				
(12.3) RAW STOCK COSTS				
(12.4) DUBBING (PRODUCTION)				
(12.5) SOUND RECORDING				
(12.6) SHIPPING AND RECEIVING				
(12.7) OTHER LIBRARY				
(13) PERSONNEL				
(13.1) PARTICIPANTS				
(13.2) WRITER				
(13.3) DIRECTOR				
(13.4) CAMERAMEN				
(13.5) OTHER PRODUCTION PERSONNEL				
(13.6) ELECTRICIAN				
(13.7) PROFESSIONAL TALENT				
(14) TRAVEL				
(14.1) TRANSPORTATION EXPENSE				
(14.2) LODGING				
(14.3) MEALS				
(14.4) MISCELLANEOUS				
(15) SPECIAL NOTES				
(16) APPROVED ____________ TOTAL, THIS PAGE:				
TOTAL, PAGE 1:				
APPROVED ____________ GRAND TOTAL:				

A variation of this is to use the indicating letter, followed by the last two digits of the year, followed by the production number. In other words: M-72-8 could represent the eighth show produced for marketing in 1972.

3. **Title.** This may seem like an obvious enough entry, but considerable thought should be devoted to the name of the show being produced. There is often a tendency to produce titles that are too long to be used comfortably (by a narrator or participant when the title is spoken, or in print form when used as a title on the screen). Keep titles short, and directly related to the subject matter.

4. **Requested by.** It is important to make note of the requester, since it is through that person that information is gathered and necessary approvals are given. It is a good idea, in most cases, to ask for the name of an alternate, so that production efforts can continue in the unexpected absence of the original requester.

5. **Description of subject.** This entry is more for historical purposes than anything else and reflects something about the true content of the film — what it's really about, or what it's supposed to do.

6. **Estimated running time.** This is only a rough guess, and not something to magically govern the length of the program. As we mention elsewhere, the nature of the story itself often governs the length of time it takes to tell the story — and is not, therefore, something that can (or necessarily should) be estimated ahead of time. Rough guesses about length are of particular importance when trying to plan for the use of outside facilities (where rental costs are an important factor). It is of less consequence in your own studio, where you can more easily juggle schedules.

7. **Indoctrination required.** This somewhat sinister heading serves only to indicate whether or not the proposed participants have ever had any experience with your studio, or with television generally. It is useful if you can explain to them what your studio does and how it operates. It will provide the participants with useful information, and make them more familiar (and more comfortable) with the medium. In the chapter on script preparation (Chapter 14) we have suggested material which might be included in a program to familiarize people with your facilities and methods. Indoctrination may consist of merely showing the prospective participant that suggested video-taped show.

8. **Time, dollars, and date.** The columns on the right-hand side of the sheet have been designed to accumulate necessary information

about the amount of time projected for various activities, the amount of money each will cost, and the dates by which these activities are to be completed. As more experience is gained this estimating job becomes easier (and the accuracy gets better).

9. **Production.** Generally speaking, there are three principal places where production effort can take place: in your own studio, on location (with your own crew), or in outside facilities (with outside production personnel).

9.1 **In studio.** An entry here indicates that the major part of the production effort will take place within your own facilities. This is important to know, since it is easier to set your own schedules for space, time, and people than it is to deal with the planning involved in location work, or in scheduling outside facilities. There is more flexibility regarding absolute deadlines in your own studio.

9.2 **On location (in-house, off-site).** This is to serve as an indication that some, or all, of the production will be shot outside the studio (but with your equipment and people). Production of this type may take place in laboratories, test facilities, production areas, distant offices, and so forth, and will entail the transportation of TV equipment, lighting units, and other materials. In some cases, rented equipment must be contracted for. Chapter 15 of this book is devoted to the special problems and techniques of location shooting.

9.3 **Outside facilities.** For certain types of production activity you may find it more sensible, convenient, or economical to use an outside production facility. It must be kept in mind that this involves some very careful schedule making and, more importantly, schedule *keeping*. Chapter 16 of this book is devoted to this subject.

10. **Equipment/power requirements.** The four entries in this category are meant to be handled with a check mark — and this will quickly serve to indicate any special needs in terms of cameras, lighting units, or other equipment. In special cases it may be necessary to mark more than one entry.

10.1 **Equipment in studio.** In most cases this entry will be checked, and this indicates that the normal complement of studio equipment will be used.

10.2 **Portable equipment.** This entry will be checked at times when it is necessary to shoot on location. If work of this type is done on a routine basis it would be a good idea to put together a standard equipment package to be used in every case when shooting is done outside the studio. Such a package would include portable camera and lighting equipment, and all necessary stands, microphones, cables, and other materials. The notation "full equipment package needed" (or some variation of this idea) could then be made in conjunction with this entry.

10.3 Rental equipment. This notation is made when equipment is required for in-studio or on-location work but does not apply to outside facility production (in which the entire operation is rented). There are numerous occasions when a specialized piece of equipment may be needed on a one-time basis. It must be remembered to arrange for rental equipment well in advance of its intended use.

10.4 Electrical power needs. When shooting on location there is always a need for suitable light levels, and this means the presence of lights, sometimes in large numbers. It is important to make the necessary plans and arrangements so that adequate power will be made available.

11. Preproduction considerations. Following are eleven key areas which must be considered prior to the actual studio work.

11.1 Production planning. The purpose of the Production Planning Sheet is to aid in the overall planning of any given project. This entry is made in order for you to set aside a certain amount of time to consider all the various production aspects (pre-, during , and post-) involved. This is vital, both from a standpoint of estimating costs, and for scheduling time.

11.2 Script or outline preparation. Small-studio productions seldom require the preparation of a complete, detailed script, but all of them require an outline. It is imperative that the information be organized intelligently, and likewise important that the established organization be understood by the production crew (so that cameras, visuals, and participants are in the right place at the right time).

11.3 Typing, reproduction, messengers. Under this category must be included all of the office, clerical, and administrative costs of copy preparation and distribution. This is usually of considerable importance, but it is apt to be overlooked. We usually forget that it costs several cents every time we make a copy on an office copying machine. This factor alone contributes considerably to the copy production expense, since multiple copies must be made of all written material.

11.4 Production of visuals (time/materials). (As in other important activities, a chapter — Chapter 10 — is devoted to the production of visual aids.) Experience is the only guide to what it costs to produce visual materials on a straight-time or overtime basis, or, for that matter, through an outside supplier. Visual aids represent, however, a very measurable part of the production costs, and must be estimated carefully. In addition, sufficient lead times must be allowed so that the visuals are on hand when required.

11.5 Photography. In some cases photographic services will be required, for materials to be used on camera, for slides to be used in a film chain, for in-studio photos during a regular shooting session, and for various other applications. Photographers (either in-house or outside) should be given as much lead time as possible.

11.6 Rehearsal or walk-through. This entry serves to remind you that this activity not only needs to be performed, but that it requires that certain studio times be set aside, and everyone connected with the production notified (and their presence assured).

11.7 Availability of tape, etc. One should routinely check to see that adequate stocks of regularly used materials are on hand. Production efforts have been stopped in the past because of the absence of some seemingly trivial item.

11.8 Sets; 11.9 Props; 11.10 Costumes; and 11.11 Makeup. These elements must be considered, and necessary decisions made, well in advance of the production effort.

12. Postproduction requirements. The following seven entries refer to activity that takes place after the studio phase of operations.

12.1 Editing. This is another of those areas in which experience is the best guide for estimating the amount of time needed to do an effective job on a particular show. The need for editing is partly a function of how well a show has been planned in the first place. If a production has been well organized and rehearsed, the chances of much editing being needed are negligible. A show that is largely unplanned (with people who are ill-prepared) will require extensive editing. It costs considerably less, in the long run, to insist that certain levels of quality be met in the studio phases of operations.

12.2 Number of dubs (duplicates). This entry requires only the numerical count of dubs needed for distribution.

12.3 Raw stock costs. This is simply a mathematical calculation of the cost of tapes to make the required number of dubs.

12.4 Dubbing (production). Based on the number of dubs required, it will be possible to estimate the amount of time (therefore the costs) for producing the dubs, either on an hourly basis using in-house capabilities or on a per-unit cost outside.

12.5 Sound recording. This includes the time and materials required to record additional dialogue, music, or sound effects on the additional sound track of the master video tape. Costs will vary according to the complexity of the sound material to be incorporated.

12.6 Shipping and receiving. If tapes are to be distributed there will be a significant cost in the preparation of the materials for shipping. Costs must be calculated in terms of time needed to do the job, packaging requirements (containers), and postage or freight.

12.7 Other library. This entry relates to the simple handling of materials, such as that required in shipping and receiving. The time of a tape librarian must be estimated, because the job of filing and cataloguing materials is an important one.

13. Personnel. There are a variety of people who may become involved in a video tape production. Their efforts must be directed and their time scheduled.

13.1 Participants. This serves as a note to the identity of the key performers in the production — where they are located and how they can be reached to gain information, or to pass along instructions.

13.2 Writer. If a script or detailed outline for a show is required, it will be necessary to assign a writer to the project, to allot him a reasonable amount of time, and to coordinate the necessary information contacts.

13.3 Director. This entry is made to show who has been given the assignment, and the production authority, and is also used to estimate the approximate amount of time that he or she will be needed — in rehearsals, walk-throughs, or in actual production.

13.4 Cameramen. This entry, once again, is made to show assignments, and for the purpose of scheduling time. In some situations, they have alternate duties, and their time must be arranged for well in advance.

13.5 Other production personnel; 13.6 Electrician; and 13.7 Professional talent. These entries are made for purposes of estimating time availability and computing costs, but a special note should be made with reference to professional talent (narrators, actors, etc.). Because they are often paid by the hour, it is well to be prepared when they arrive.

14. Travel. In this four-part section are the component parts needed to estimate travel costs. They needn't be gone into in any detail. Suffice it to say that extensive on-location production, particularly at considerable distances from the studio, can represent a major expense, and must be estimated with care.

15. Special notes. Any unusual requirements of the production can be entered in the space provided.

16. Approved. Spaces have been provided for the signatures of any necessary authorizing personnel.

CHAPTER 7

THE PERFORMER

The first consideration in selecting a performer to appear in your production is the choice of an individual who has the most knowledge of the subject, or one whose position carries the authority of knowledge. Such considerations as "Will he or she have television or audience appeal?" can only be determined after an on-camera test, or after that person has received a bit of TV coaching. In the end, there is no substitute for the one who knows the subject and speaks with the assurance of an expert.

Likewise, it is advisable to bring in a series of experts to carry the individual segments of a total program. For example, a marketing plan might best be handled by the marketing director, but at the point of describing the product, the chief engineer may be called upon. It may also be desirable to have the president of a company say a few words to indicate that top management is truly behind the marketing drive.

In other words, the television screen must continue to reinforce the believability that the audience associates with the medium. There are times, of course, when the most authoritative person clearly lacks the personality or the "presence" required to get a message across effectively. The simplest solution in this case is to find another performer.

There are also times you will require the services of a professional performer, especially when there is need for an off-camera announcer or narrator. If an audience cannot *see* (or identify) a speaker, then it expects to *hear* a professional delivery.

Television is a very intimate medium. Like the motion picture, it is possible for the camera to get close to the speaker and make him appear larger than life-size on the screen. Thus, television is also a very revealing medium. When a national figure appears on "Meet The Press," for example, the audience has an opportunity to assess the guest's personality and sincerity. There is no distance between audience and performer — there is no place to hide — and there is no way to conceal uneasiness.

Therefore, probably the most important rule for the performer to remember is: to be oneself. Attempting to imitate a professional performer whom one admires for a smooth delivery is a dangerous game to play. It can only appear artificial to the viewing audience.

If we were attempting to oversimplify, we could say "forget you're on television and act naturally," but this, of course, is silly. There is nothing natural about the studio situation, with its cameras, lights, settings, and restrictions. Continued familiarity with this environment, however, can lead the performer to grow accustomed to it, and to learn how to deal more effectively with it.

It would seem that one way to get an inexperienced person through a television performance would be to have him or her depend solely on a script — to read it and never depart from it.

A performer *can* read a script, but the effectiveness will be lessened considerably because there will be no contact with the audience. The audience would merely be viewing a person reading a script, instead of having the ability to look at the person's facial expressions and measure sincerity and enthusiasm.

If a performer must rely upon a script, the producer or director should first consider substituting another performer. If that is not an alternative, then the purchase of a prompter might be considered.

There are two basic types of mechanical prompters. The older type utilizes a paper roll that slowly unwinds and reveals the script. It can be attached directly to the camera, or can be placed on a stand if that is more convenient for the performer. A special, large-faced typewriter is required for the preparation of these speech rolls. The disadvantage of this type of prompter is that the performer can never look directly into the camera lens and establish direct contact with the audience.

The more recent prompters operate with a two-way mirror whereby the image of the script is reflected off one side of a pane of glass (while the camera is shooting directly through the other side of the glass). Thus, the speaker is looking directly into the camera lens while reading the script off the prompter.

A technician controls the pace of the prompter with the rate of delivery of the speaker. A speaker who wishes to depart from the prepared script may do so, and the prompter will wait until the speaker returns to the prepared copy.

Most people who appear in small-studio productions are not comfortable in the glare of the lights. They have a tendency to be restricted in their thoughts and delivery, and a prompter provides them needed help. On the other hand, most amateur performers are poor readers, and presenting them with a full script prepared on a prompter usually results in a stilted delivery.

A simpler method for small-studio use is cue cards, sometimes called *idiot* cards. These are simply large pieces of poster board upon which has been written (usually with 1/4-inch black markers) the key information such as the major numerical information of a speech, certain lines of a dialogue, and so forth.

In a formal presentation (such as that of a company president delivering an annual message), a full script is probably called for, provided there are adequate rehearsals to assure smooth delivery. If a script *must* be read, then no effort should be made to hide this fact. It will only result in a choppy delivery. In most other instances, however, the performer should be provided only a key-word or key-phrase outline that will serve to keep him or her on the track of the information being presented. In every case, it is better and more believable to paraphrase, or speak extemporaneously, than it is to read a script or to recite something that has been memorized.

The timing of things must be worked out in advance and, of course, rehearsed. After the first run-through, the performer should then ignore the problem of time and leave it to the director. The director will indicate with special signals if it is necessary to pick up, or slow down, the pace — or to indicate when the program is intended to end. Since small-studio productions need not meet severe time limitations (as network programming does, for example), the element of time is far less precise. The main concern is to hold the audience's attention.

In a standard classroom or auditorium situation, the two main criteria are that a speaker talk in a voice loud enough to be heard, and distinct enough to be understood. Other refinements or embellishments in speech technique are primarily for emphasis, attention, and interest. When appearing on television, the requirement for volume can be disregarded because microphones are present and can be adjusted to pick up the voice of the speaker. Therefore, the first consideration in an effective television presentation is the quality of the voice with regard to articulation, tone, and pace.

The voice should have a natural flow, a conversational tone, and should be accompanied by gestures and movements natural to the performer.

Bearing in mind that television is a visual medium, it is not necessary to keep up a constant stream of words to the camera in order to hold attention. A simple movement such as walking from a desk to a blackboard is an attention-holding technique. The simple procedure of looking from the camera to one's notes, in order to pick up the next thought, is also an attention-holding device. The audience is patient, and will wait for words when they can see what is going on. They can see the *reason* and the naturalness of the delay, however minor.

In front of a group of people in any other presentation situation, it is sometimes necessary to exaggerate in order to communicate successfully. A gesture, for example, must be broad. In television, or film production, however, the broad gesture must be eliminated. In television, small gestures project easily and understandably. A quizzical expression, a frown — these communicate directly. In other words, subtlety is the key.

A gesture that one should consciously avoid is any hand motion near the face, such as resting one's chin in the palm of one's hand, or covering one's mouth.

One of the areas that should be established between the director and the performer is where the focal point or the attention of the performer will be directed when he or she is speaking. This, of course, not only helps the performer, but also assists the director in setting the cameras. It is not always necessary to direct attention into the lens of the camera. If there is more than one participant in the show, then it is quite natural for the attention of the participants to be directed toward the person speaking.

And in this particular situation, it is only necessary to speak to the camera at the opening and closing of the discussion, or if a moderator wishes to change the emphasis from the people to graphics or an object, etc.

Primarily, however, the problem of visual attention is the director's headache, and the performer should merely be concerned with getting the information across and not necessarily with how it is going to look on the screen.

The more experience the performer has, the greater will be his or her ability to "play to" the camera.

In addition to being visually revealing, television is aurally revealing. Whispers and other extraneous noises are picked up on the sensitive microphones. A performer should be cautioned to speak with slightly less than normal volume, to avoid drumming on desk tops or lecterns with fingers or hammering with pencils, etc. This makes a tremendous racket on the microphones and causes a considerable distraction to the audience. One must also be careful about the scraping of chairs.

As has been indicated, in a television production, each physical movement (where people change position or relationship to objects or other people) must be planned and rehearsed beforehand. The assumption is that the personnel working with cameras usually are not well enough trained to anticipate sudden, unexpected movement. Nor are they familiar enough with the equipment to make necessary lens adjustments to maintain focus and keep people composed properly in the picture.

Unexpected movements may require that the director constantly use long shots of the participants, which have less appeal than the closeups of facial expressions.

There are no set rules about clothing on television, except that extremes in contrast should generally be avoided, since contrasts in color and value have a tendency to "bloom" somewhat, detracting from the performer. Solid colors are best advised for both men and women: medium-shade suits, dark ties, and pastel-colored shirts.

When two or more people appear on a show, it might be well to advise them all to wear gray or dark suits to avoid unpleasant contrasts of colors or design. It should be remembered that light colors attract more attention than dark ones. Thus, light-colored suits may distract attention from a speaker's face. The other reason for darker suits is to provide the needed contrast for superimposition cards or slides that provide information such as name and title. Women who appear on the show should be cautioned not to wear jewelry that will reflect studio lights, causing sharp, distracting highlights in the picture.

Makeup is not only desirable, but in some instances altogether necessary. It is applied to correct certain problems that the camera would tend to emphasize — a shiny forehead, a heavy beard, etc. Studio lights bouncing off a high, shiny forehead can cause an unfortunate distraction from an otherwise excellent presentation. Makeup should never be applied so heavily that it changes the appearance of the subject. Men usually require only a pancake makeup followed by a light application of powder. Women usually need only some lipstick and eyeliner, in addition to their normal street makeup.

We have found that rehearsals can be speeded up and delays can be avoided if performers are advised beforehand of the preparation needed to appear on television.

A letter such as the following should be sent to persons scheduled to appear on a show.

SO YOU'RE GOING TO APPEAR ON TELEVISION

If this is your first appearance on television, this letter will serve to answer some of the questions you may have. If you are familiar with television production, we hope this will refresh your memory on a few essential things that must be kept in mind.

Program Content

Ours is a discussion show and maintains its interest by the give-and-take between the participants. But despite its freewheeling discussion, there is need for a few structured questions and answers in order to get the show started and to assure that the important topics are adequately covered. Therefore, please come prepared with a few notes to remind you of how you wish to phrase certain questions or answers.

Notes

If you wish to refer to notes during the course of the show, please prepare them on 3″ × 5″ cards that can be held in the hand. Please do not prepare them on standard paper. (Paper noise is easily picked up by microphones, and white sheets of paper give our cameras problems.)

Visual Aids

If you plan to use visual aids, please check with us well in advance of the production date. We will advise you on the most effective way to visually present your information; the amount of information that can be presented on a slide or chart; and we will assist you in getting the visuals made so they may be used in the television production.

Delivery

As we stated, ours is a discussion program, and your remarks should be directed to the other participants in the show. The only time we request you look at the camera is at the opening and closing of the show, when the moderator is making appropriate introductions or thanking you for your participation.

Also, you will have no way of knowing whether or not you are on-camera (even if you are not speaking, you may still be in the picture), so always keep your attention focused on the other participants. Do not get distracted by the production activity going on in the studio.

Do not be concerned about mistakes.

If you make an error or wish to rephrase something, continue with the discussion and try to verbally correct it. If you cannot, then still continue with the show and corrective action will be taken later. Do not stop the program until the director announces in the studio "cut."

Please remain in your chair after the show is over until the director announces in the studio that the tape is no longer recording.

Microphones

Speak with your normal voice at conversational level. Any problems in being heard will be corrected in the control room. Also, remember that microphones pick up *everything* — including coughs, sneezes, and whispers.

Clothing and Makeup

Generally you may wear what you wish, although it is best to stay away from stripes or plaids. Dark clothes are preferred because they set off the face better, and provide a background for names when they are superimposed on the screen.

Colored shirts are preferred to white ones although not absolutely required. Light blue is the usual color.

Over-the-calf socks are recommended for the men.

Reflective jewelry, such as rhinestone pins or earrings, are *not* recommended for the ladies.

Street makeup is recommended for the ladies. Any required base or powder will be applied at the studio — for both ladies and men.

Above all, please keep in mind that you have been asked to appear on our program because our audience wishes to see and hear you. So relax . . . be natural . . . and enjoy the experience.

CHAPTER 8

TV DIRECTION

As was stated in Chapter 5 ("Production Personnel") the key figure in the actual production process is the director — the person who is responsible for translating a production idea into a final program that can be telecast or taped, and then overseeing all the production elements during the actual run of the show.

The director operates in three major areas: (1) in the conference room and studio (or on location) *before* the production begins; (2) in the control booth *during* production; and (3) in the editing room *after* production. We will examine his or her role in each of these areas.

There are two schools of thought regarding the selection of a director. One thought is that person should be most knowledgeable in the *use* of the video equipment. He or she does not need to be a video technician but must be familiar with the capabilities and limitations of the medium. Having mastered the medium, the director is then free to learn as much as possible about the subject matter and apply the one knowledge to the other.

The second thought is to get someone knowledgeable in the subject who can assist and edit the content of the program. That person can then learn the technical aspects of the medium, but his strength will lie in the information presented.

We are inclined to go along with the first reasoning — find a good craftsman who is willing to study the material being presented. Our rationale is that we believe, for example, that the "finest" technical writing is performed by professional writers who can translate technical material with creative flair. There are notable exceptions to this, but they are still exceptions. In the same manner, the ability to use a medium creatively, such as television, far outweighs a pedantic approach no matter how much better is the content. If the content is not well received by the audience, then the effort is wasted. Naturally, if a professional industrial or educational person shows creative flair and technical interest in television, that person should be given the opportunity to direct the operation.

The good television director is, indeed, a rare combination of technician and creative innovator. He or she must possess a broad background in all creative art forms (such as painting, music, photography) and must not only be skilled in seeing and hearing good art, but also be able to determine *why* it is good and how it can apply to a particular directorial assignment. This "sense of art" is most difficult to teach. It can be brought out of a person, but cannot be instilled.

In the same manner, technology frightens many creative persons. They choose either to ignore physical limitations of certain equipment, or refuse to take the time to investigate the broad range of the equipment's creative capabilities. And they sometimes get so carried away with creative flair that they fail to account for the purpose of the production and audience participation and reaction.

This is not to suggest that good television directors are not available — there are many who can work effectively in an industrial or educational situation. But you may find that the environment of the small studio, and the specific audience programming, will require compromises in both selection of the director and that person's ability to be as creative as he or she wishes.

One important aspect of direction is the ability to work with people and to make quick and decisive judgments. The conditions of the very situation, with novice persons working under tight scheduling on something that will be critically viewed, do not permit much room for temperamental displays or abrasive personalities.

In addition, the final decision-making authority during the actual production rests with the director. If he or she repeatedly makes changes, or in any manner loses the confidence of the technical crew and talent, each person will tend to seek individual solutions and the result will be chaos.

The director's first job, therefore, is to *plan* the production. He or she is dealing with ideas and people — especially people. He or she will probably be working with performers whose main professional function is in other areas and of more importance to them than the television production. As little of their time as possible must be wasted, and when they show up at the studio or on location they must be convinced that the program is as well planned as possible. The cameras, lights, and other technical aspects of production must be ready and rehearsed by the time the performers enter the set.

The director, therefore, must carefully plan the staging and the camera shots, and know exactly what is going to happen during the show. Certainly there will be times when a director has to "wing it," but they should be rare.

The program is conceived when someone decides on a need to communicate some information to an audience. The program is produced when one individual is assigned the responsibility of production. (Committees don't produce programming.) That individual calls in a director and explains the purpose of the show and to whom it is being presented.

One of the first decisions to be made by the producer and director is the format of the show. Will it be a discussion program, an interview, a how-to-do-it demonstration?

Next, there is the determination of *where* it will happen — where the scene will take place. (A *scene* is a single location that may comprise the entire show, or be one of many scenes that comprise the show. One or more *shots* make a scene; one or more *scenes* make a show.)

Once the format and location of the program have been determined, the director must think in terms of pictures, and sequences of pictures. He or she must think in terms of uncluttered shots, since each picture will be confined to the small size of a television screen. The medium is too small to communicate an abundance of information in a single shot. Mass information must be broken down into viewable bits that are properly sequenced and visually appealing. Also, extremely fine detail is going to get lost in the system, and you may have to present a broader viewpoint than desired.

The first element with which the director must be concerned is that of the setting. Will the background look like the television studio? An office? A classroom? Will background objects be defined, or will everything take place in front of drapes or no-seam paper? The actual setting will be worked out with the designer (who also may be the director), but the preproduction planning must consider the amount of space the setting will use, the placement of the objects and the people, and the available space left over for positioning of the cameras.

A detailed sketch of the studio and the space that will be utilized must be drawn by the director to ensure that everything will fit, and that there will be enough room for the cameras and equipment. The director must position the cameras so they will be able to capture all the action with the least amount of movement,

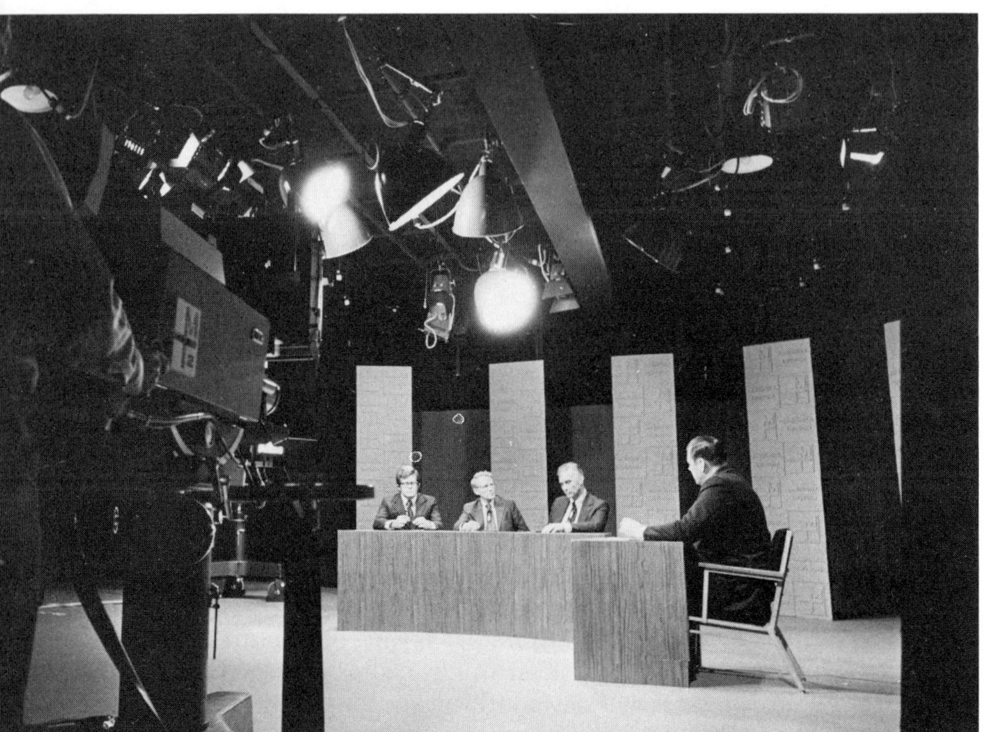

(Photograph courtesy Merrill Lynch, Pierce, Fenner & Smith, Inc.)

Simple staging with an interesting background is recommended for small-studio productions

and at the same time allow for varying and pleasing angles. In the same manner he or she must plot the movement of the performers so they stay within the range of the cameras and do not cause angle or position problems.

The director need not be concerned about the actual shots that one envisions, because, once inside the studio, the need for changes may be apparent from seeing how certain things actually appear on the screen. But a general attitude regarding positions, angles, and heights must be considered using the preproduction sketch.

The director must also keep in mind that television has the ability to integrate other communications media: artwork, charts, slides, photos, film, music, sound effects. He or she must remember that the medium is primarily visual and it must move — and motion comes from the movement of people, lenses, cameras.

It is tough enough to ask an audience to view a single person on camera for any length of time. But then to also ask them to watch while that person speaks head-on into a camera that never changes its position — this is cause for channel-switching. The early days of educational television almost caused their own death knell because ETV producers tried to move the classroom to the studio, instead of adapting the classroom to the medium.

In a single-camera system, you are fairly confined to movement by the person and the changing of focal length of the lens (zooming). Most 1-inch and 1/2-inch equipment is too lightweight for practical changes of camera position during the actual recording of the show. The movement is manual and must be performed by the cameraman while trying to keep the subject in frame and in focus. Effective dollying and trucking of cameras can only be achieved with heavy commercial broadcast tripods, and sometimes with the assistance of another man.

It is best, therefore, to keep the movements and shots relatively simple, especially during the first trial productions. There will be ample time to experiment and add interesting production techniques after you have mastered the basic capability and intent of the studio.

Now if it all appears to work on paper, then you're ready to move into the studio.

The fledgling director must resist the temptation to overdirect, i.e., make too many changes in angles, fades, dissolves. It must be kept in mind that a television production — like any communication medium — has *pace*. That is, part of the effectiveness or impact of the message depends upon how well the show *moves*. An overabundance of movement without regard for content or impact is just as bad as no movement at all. For example, to change camera angles during the time the speaker is stating the main point of the presentation may be distracting and mean loss of impact.

Therefore, the first requirement of the director in the control booth is to become acquainted with the production techniques available and have some knowledge of when they should be used.

One of the key things to remember about the technology of television is its ability to produce electronically certain kinds of visual effects — visual effects that are very hard to come by in motion picture production.

In film production, for example, a fade-in or a fade-out is an effect that can be done in a camera, but is usually done in the printer at a film-processing laboratory. The printer gate is programmed to open and close at a prescribed rate so that the length of the fade is precisely measured. The finished film is then processed. A dissolve is more complicated, since it consists of essentially two fades taking place simultaneously — a fade-out of one scene superimposed on the fade-in from another scene.

This activity in a laboratory must be done with a great deal of precision. In television production it is a simple matter of slowly moving the fader bar on the control panel.

The main purpose of a visual effect *within* a scene is for variety. The purpose of a visual effect *between* scenes is for transition.

If you are using one camera the most basic change in a picture is your relative closeness to a person or an object. This relationship (how close, or how far back you are from the subject) is performed with a zoom lens. And movement should occur frequently, but with a purpose (a close-up for emphasis . . . a medium shot for transition).

In a multicamera situation, the basic change between cameras is achieved by a "cut." A cut is instantaneous and implies no change in time. You can cut between cameras if you wish to pick up the action from another angle or location. It is not advisable to cut between cameras while either camera is in the process of a zoom. However, it is advisable to cut while a performer is moving, so that as a figure gets into an awkward angle with one camera, he or she is picked up with a better angle by another camera.

A "dissolve" effect usually implies no change in time, but a change in location (or, "Meanwhile, back at the ranch . . . ."). However, some dissolves are used effectively on a single speaker when making a transition from one extreme angle to another (a long shot from head-on to an extreme close-up of a side view).

A "fade" is used between scenes. A fade-in gets the scene started and a fade-out concludes it. The speed of a fade depends upon the overall pace of the show, and this a director must sense. A type of fade is a "wipe" but this should be used sparingly.

Other visual effects are split screens, corner wipes, superimpositions, and whatever else your switcher/fader is provided with. As we have indicated before, these are production techniques which should be utilized after mastering the basics. Their value can only be measured by actually viewing the results on the screen to see if you have achieved the proper impact without gimmickry or distractions.

There is a great deal that can be done with camera technique to create a professional-looking job. The main way to achieve desired results is to have the camera in position, focused on the subject, with the picture composed before the switch to that camera is made.

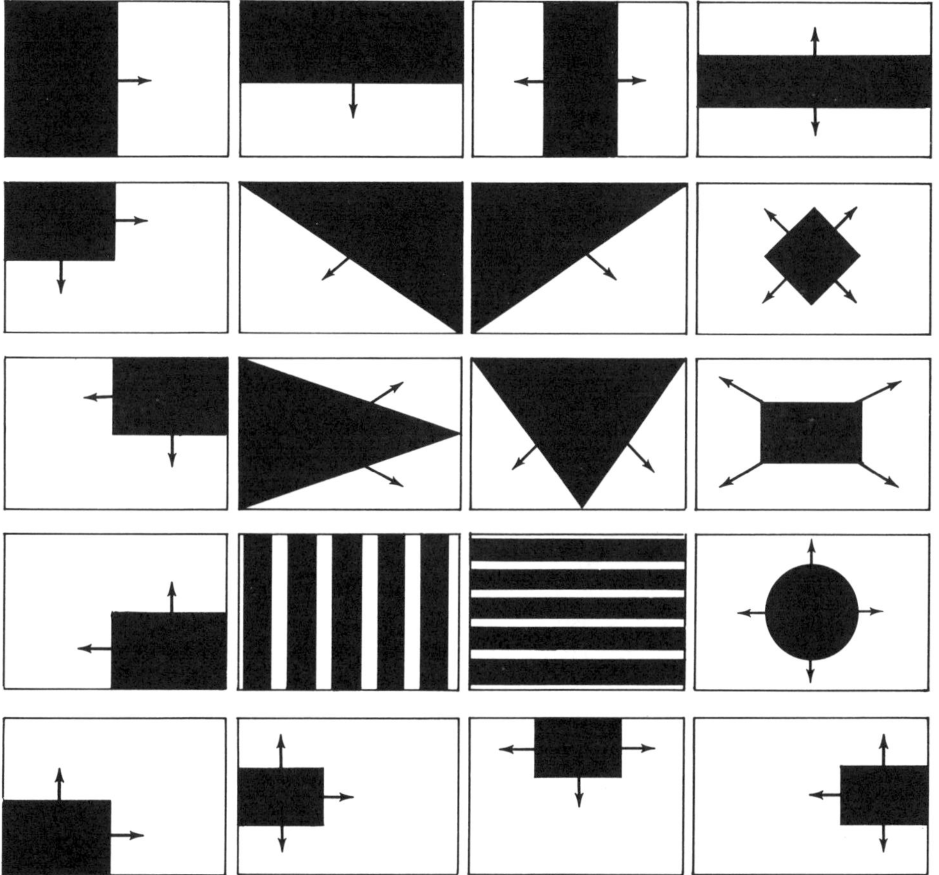

(From Producers Manual, Magnetic Products Division, 3M Company)

Wipes possible with special-effects generator

The small-studio director will probably do the actual switching in the control room. He or she will also be looking at camera monitors, the line monitor, and, if available, a preview monitor for slides and charts. At the same time the director must listen to what the performers are saying in order to anticipate movement and cues for projection of other material. Therefore the director must give clear and concise instructions to the other production personnel to maintain the proper pace and avoid extraneous movement.

The director can use whatever terms he or she desires, but whatever the terms, their purpose is to tell the camera operators as quickly as possible upon whom their cameras should be focused and the type of shot desired. In most business situations, the person appearing on the screen will probably be known to those operating the cameras, and the simple expedient of using the individual's name may be the easiest system to use. The number of individuals to be included in a shot is also quite simply described: it is either a one-shot, a two-shot, or a three-shot, etc. If the total group is to be included it is called a group-shot.

The composition of pictures is a matter of judgment and taste. The first requisite is that the viewer be able to see what is going on. From that point, the shot will be only as interesting as the cameraman is capable. The director can point out little refinements, such as putting a face slightly off-center of the frame when the speaker is looking off-camera. But the speed of switching and integrating technical aspects usually precludes the director's composing a picture other than determining who is to appear in the picture, from what angle, and to what degree of closeness.

When dealing with a live performance, the director must develop a sense of anticipating what will happen. Even with a single speaker, the director must sense the performer's speed of delivery, gestures, and movements. By getting the pace of the performer, the director can also set a pace for changing lens positions and camera angles.

When directing a panel discussion, the director must try to anticipate who will next respond. The more cameras used the more the opportunities of picking up the speaker. Even so, the director has numerous options available so the audience will not miss who is speaking. One option is to have one camera continually set on a long shot of the entire group, with the remaining camera(s) picking up the individual speakers. As one speaker stops, it is possible to switch to the long shot until the second camera has picked up and focused on the succeeding speaker.

Another option is to divide the group in half, with one camera covering one half and the second camera covering the other. (A third camera can cover the moderator or free-wheel.) As a speaker ends his or her remarks, the camera operator zooms back to cover half of the group. Depending on who speaks next, the director selects a camera, and the camera zooms in for a close-up. If the comments are coming short and fast, the cameras merely stay on the long shot, and possibly pan from speaker to speaker.

Whatever system is used, some of the opening shots on the show must be used to establish where the people are located. The opening shot can be a long shot of the entire group and the setting. Or, the opening shot can be a close-up of the main guest, with the lens pulling back to show the physical relationship of that individual to the other participants.

As we indicated, any system can be used to notify the other production personnel of what is wanted. But the system should consist of: (1) instructions, (2) a warning, and (3) use. For example, one could say, "Camera 2, get me a two-shot of Smith and Jones. Camera 3, I need a close-up of the moderator. Stand by camera 2. Cut to camera 2 . . . *now*. Camera 1, a long shot of the group. Stand by camera 3. Dissolve to camera 3 . . . *now*."

Following is a description of a typical interview-type show, with all aspects of the directorial effort described.

The show opens with camera 1 on an easel board that is at the same level as the camera lens. The card reads: "A Marketing Forecast for the 1980s." The card also shows the logotype of the company producing the information, and the division of the organization.

Camera 2 is on the set, and has a medium shot of the first speaker, in this case the interviewer, who is seated at a table. Behind the speaker is a blackboard which is on the wall of an office setting. The camera is focused specifically on the first speaker.

Just prior to the actual taping the director indicates to the talent that production is about to begin, and brings the VTR up to speed. Signal lights above the studio doors indicate that taping is in progress, and that silence on the set is essential.

The cameramen are told to "stand by" and the director takes the first shot.

The director fades in on camera 1 and the title "A Marketing Forecast for the 1980s." The console monitor shows the director that camera 2 is on the opening medium shot of the speaker at the table.

The director says "Stand by, camera 2 . . . *now*," at which time he or she "takes" that camera and switches from 1 to 2.

The speaker begins with a rehearsal opening such as:

> *Hello, I'm Bob Johnson of the marketing department. I know that a lot of the salesmen in the field have said, from time to time, that they wish they had a clearer idea of what the marketing plans of the company were, and the ways marketing strategies were developed.*

By this time the director has camera 1 break from its shot of the title and take a different shot (different angle and different closeness of shot) of the speaker. The operator of camera number 1 has been made responsible for all the artwork pickup shots, and has rehearsed them ahead of time.

At this time, camera 1 has gone in for a head-and-shoulders of the speaker and the director can see it on the camera monitor. The director says "Stand by, camera 1" (thus alerting the operator to lock the camera in position and to stay on the shot) " . . . *now*." The red light on camera 1 is now on and the talent can direct his or her attention to that lens.

"O.K., camera 2, get a two-shot of the speaker and his guest."

The narration continues:

> *In order to give you a clearer view of how marketing data is acquired, and how it is used in long-range programs, I have asked one of the consultants of our company, an economics professor, to describe, in a general way, our markets.*

The director now says "O.K., camera 2, stand by on your two-shot . . . *now*," and switches from the tight shot of the speaker to a medium shot of both men at the table.

> *This is Mr. Harold Hellman. Harold has worked with the marketing division for two years now, and has some insights into the problem that you will find of interest.*

The director asks camera 1 to go for a fairly tight shot of the consultant, so that when he begins speaking, the camera will be on him only.

> *Harold, what are the benefits we accrue from looking at the problem in a much larger perspective — in other words, an extremely long-range forecast?*

"Stand by, camera 1, . . . *now*."

The consultant looks directly into the camera and says:

> *Well, all too often when we take the short view, the day-to-day impression of the market and what we think it looks like, we can't perceive trends. We can only begin to tell something about the future when we look not only at our own past experience and all the data that comes from that, but also at other things that may have an influence on the market in five or ten years time.*

The director's notes indicate that very early in the presentation the consultant will use the blackboard to begin the visualization of an idea, and then cut to a piece of finished art in which the entire concept is revealed. Aware that camera 1 must break for the artwork, the director therefore calls for camera 2 to go for a wide shot of the two men and the blackboard. In this way, the camera won't have to try to follow the action of a person while it is on a tight shot. To follow somebody on a tight lens is practically impossible and the only time it should be done is when the action or the mood of the story calls for it. (In other words it is perfectly acceptable when following the action of a man in a drama in which, let's say, he's trying to escape from someone . . . the erratic motion of the tight shot would then help convey some of the tenseness and excitement of both the man and the situation.) There is no need for this kind of effect in an interview, however, so we want the visual part of the story to flow as naturally as possible.

Camera 2 finds the wide shot in which it will not be necessary to tilt the camera when the man stands up. The director "takes" the shot, and frees camera 1 to set up on the artwork.

> *Let's look at a growth curve for a product.*

The consultant stands and goes to the board. He draws a rectangle in which he places a bell curve. Once he has established himself at the board it may be possible for the director to have camera 2, the live camera, zoom in for a somewhat tighter picture of the man at the board.

> *At the front of this curve we have a situation in which the new product is being introduced to the market. The very first of this cycle can be seen in greater detail in another drawing.*

The director takes camera 1, which is already set up on the artwork.

The speaker then takes his seat to continue his description of the artwork. It may be that a second piece of artwork is needed to follow the first. In that case, it could be set up on another easel in line with camera 2 and located on its side of the studio. This would be the most professional way of handling it, since it would enable the director to dissolve from one shot to the other.

As the conversation continues, the two cameras continue to alternate providing both variety for the viewer and making it possible to go from photos and artwork to the speakers. At one point, it may be useful to use both cameras simultaneously, for instance, when the name of the speaker is superimposed on his live picture.

The recording of this particular session may now continue until the end, at which time an end title may be used. (Music in and out of the production may also be a requirement, in which case ample rehearsal should be made so this is done smoothly and easily. A disc or tape player can readily be handled by the director if the equipment is located sensibly. In other words, an additional person shouldn't be required in the control booth.

One thing to keep in mind is that your television show can be much more relaxed than a commercial show — not more relaxed in standards, but in performance. There is no need for hiding the fact that the show is being staged in a television studio, and that it is not fully scripted. This means that the performer can aid in the staging of the production and help anticipate when the next shot will be.

The performer can assist the director by talking ahead of the visuals: "The next shot will show . . ." or if the performer decides to insert a shot that was not planned: "If we can get a shot of that last chart again, I want to point out . . ."

It is best, of course, not to rely upon the performer to keep the show going. The glib sales manager or articulate instructor is not as relaxed in the studio as in the preproduction discussions. All nonprofessionals have a tendency to tighten up during the course of a television production. Therefore, the rehearsals with the performers are most essential so persons on both sides of the camera will be aware of what comes next.

In a discussion show, for example, the areas to be covered, and some specific questions to be asked, should be prepared and rehearsed. In this manner, the participants can get off to a distinct start. Soon the questions are fresh and cause some thought and reaction. But by this time the participants are relaxed and will be more inclined to be natural in their answers and their actions.

With most discussion shows there is a tendency for the participants to be overly polite to each other or to measure their remarks. The speakers should be encouraged to raise issues, challenge answers, and, generally, speak their minds as they would in off-camera discussion.

As we have said, for some first-time performers, the studio and lights and cameras can be overwhelming and cause apprehension and tension. One technique used to overcome this condition is to get them to rehearse, but with the emphasis that you, the director, need the rehearsal. One can go so far as to ignore the performers and speak only to the technicians. When the performers realize they aren't the only persons in the production process and that all eyes aren't pointed at them, they tend to relax so they can help the others get through.

A director may wish to record some of the rehearsals in case some later picture inserts are needed to cover some narrative fluffs. But the taped rehearsal should not be shown to the novice performer. It will only reinforce his nervousness.

Performers should be advised that despite their impression of flawless productions on commercial television, mistakes often occur but can be corrected before the viewer sees the show. However, in order to make an undetected correction, the continuity of the show must be maintained — both continuity of picture and voice. It is best, when dealing with amateur performers, to keep a sequence going rather than try to stop and pick up on a specific word or head position.

Thus, a speaker who wishes to correct something just said, instead of stopping the taping, should continue with words such as, "That isn't what I meant to say. What I really wanted to say was . . ." During those few seconds, one can switch cameras to a completely different angle or shot (from a close-up to a long shot). When it comes time to edit out the unwanted words, a change in head position or gesture will not be noticed.

It is advised to start over completely only when a mistake occurs within the first few minutes of the show. A fresh start is sometimes relaxing and generates new enthusiasm. (Too many starts dull the enthusiasm, however.)

The farther into the show a mistake happens, the less one should consider starting over from the beginning. Amateur performers are naturally nervous and errors have a tendency to multiply. If a complete run-through can be accomplished with a minimum of fluffs, one can then edit out the errors completely or insert covering shots. But continually repeating the shooting of a half-hour show — hoping for a flawless performance — may also result in a measured, unenthusiastic presentation.

With more experienced participants the director may not feel that the show needs a rehearsal, but will settle for a walk-through — a basic process of establishing the key points and key movements of the show.

"Business" represents the on-camera actions and movements of a performer, experienced or otherwise. Often, when a performer has nothing to do or say in an interval of time on-camera, the director (or the actor himself) will provide some believable activity to keep him at ease. However, business should not be the kind of activity that detracts from the performance of another participant, but should be natural-appearing actions and gestures that add to the mood and pace of the scene. The following are natural pieces of business:

1. Examining some papers (if there is motivation for doing so)
2. Gesturing for emphasis
3. Standing up and moving on the set
4. Making notes with pencil or pen (in a limited way . . . and not so as to detract from the main speaker or focus of attention).

Looking at the speaker can often represent stage business. (This is very important to do, by the way. It makes the entire scene more believable, and makes it seem that the program is really originating "live" and that the information being imparted is being heard for the first time.)

Avoid such actions as lighting cigarettes, because smoke is eye-catching and not easily dispersed in the studio. However, if a guest is generally known as a person who continually smokes a cigar or pipe, this bit of business should be incorporated.

To get a performer started during the taping, it will be necessary to work out some system of cueing. Don't depend upon the red light on the camera — it comes on so suddenly the performers are usually unprepared or startled. Have the floor manager (or one of the cameramen) give a signal to anticipate the cue (usually a raised arm), and then the actual cue (the arm coming down pointing to the performer).

To get the performer off the air, execute a signal to *wind it up* (rotating the hands one around the other as a signal to hurry); or give the signal to *cut it off* (the gesture of cutting the throat with the edge of the hand).

The final phase of directing is in the editing room. In most small-studio situations, editing is merely to correct errors or add pictures or sound that could not be obtained in the studio. Where in the production they are to be inserted, and the length of their running time, are up to the director, who must inform the tape editor accordingly.

Once the inserts have been decided upon, the cuts have been made, and the audio inserts completed, the tape editor can perform the functions and the job of the director is over.

CHAPTER 9

CAMERA TECHNIQUE

The basic story-telling device of television is the picture seen by the TV camera at any given time.

As mentioned earlier, the television camera has the advantage of instant viewing . . . and what appears on the cameraman's viewer and the director's monitor is the same picture that will be seen by the audience. The cameraman, therefore, plays a vital, creative role in the television production.

Camera operators deal with *shots* — specific positions and settings of the camera and its lenses. A shot may be held for only a few seconds, or it may last for minutes. There may be movement within the shot by either the performer or the operator, or both the camera and the subject matter may be stationary. Whatever the position or movement, the shot is designed to show what is happening visually.

The camera shot is a function of a number of things, the first of which is the choice of the particular lens used.

Lenses are measured in terms of focal length, and the larger (numerically) the focal length, the narrower the field of view. A lens with a small number will be a wide-angle lens; a lens with a large number will be a telephoto lens. Because TV lenses are mainly used in studios, there are few telephoto television lenses (the exceptions being, of course, the lenses used for sporting events).

A 75mm lens in small-studio work would be considered a "long" lens or one that had a narrow field of view. An example of a "short" lens would be a 10mm or 12mm lens, which would have a wide angle of view. A 25mm lens is considered a normal lens for a vidicon tube.

All television lenses (or good lenses of any photographic system) are equipped with iris openings which control the amount of light passing through the lens. To open up the iris allows a greater amount of light to fall into the TV camera and onto the pickup tube. To close the iris will limit the amount of light, or lessen it.

Iris settings on camera lenses are measured in terms of f-stops, and these are expressed in terms of numbers. The lower the f-stop number, the greater the opening of the iris and the greater the amount of light falling into the system. The larger the f-stop number the less light being admitted. (In going from small numbers to larger ones, each numerical progression reduces the amount of admitted light by half.) The terminology used in dealing with f-stops is to *open up* the lens or *close down* the lens. This is usually tested about one stop at a time. Remember, to open up a lens one f-stop is to effectively double the amount of light on the pickup tube. To close down one stop is to reduce the amount of light by half.

There is a direct correlation between f-stops and the all-important consideration of *depth of field.*

Depth of field is the distance in which everything in the shot is in focus, from the front to the back. A way to envision it is to consider a performer who is walking toward the camera. With a certain lens (and f-stop setting) the performer will be out of focus while walking toward the camera. At a certain point he or she will be in focus and continue to be in focus while advancing. After perhaps a few feet the performer will again be out of focus (from being too close to the lens). The area passed through was the effective depth of field of that particular shot.

Short focal-length lenses have a greater depth of field than long focal-length lenses.

Each of these factors is an important consideration in terms of lighting. The greater the amount of light, the more the camera can be stopped down (the iris closed down so as to admit less light). The greater the degree to which the lenses are stopped down means the greater the depth of field, the greater the distance at which things will be in focus.

Conversely, the less light, the larger the lens opening, and the shorter the depth of field.

This explains why, in movie and television production, it is necessary to have such a high level of light. It permits greater depth of field and the use of various lenses to their best capability.

Another significant aspect of lenses is the *field of view*, or the area seen by a particular lens and expressed in degrees. Typically speaking, a telephoto lens has a field of view of about 15 to 30 degrees. A standard lens may see 20 to 50 degrees, and a wide-angle lens about 60 to 100 degrees. (The human eye, with its peripheral vision, has a field of view of about 120 degrees.)

The final key factors in a camera shot are the lens-to-subject distance and the size of the subject being shot.

In other words, a head-and-shoulders shot of a performer can be shot with any number of lenses and the size of the head would be the same in each shot. A wide-angle lens, however, used very close to the subject would create distortions in the features, particularly in the apparent size of the performer's nose. A much longer lens would require the camera to be at the extreme end of the studio and to operate with depth-of-field problems. The best choice would be a lens that could be used conveniently (in terms of subject and distance from subject), and would not cause distortion.

For small-studio operations a zoom lens offers the greatest amount of flexibility. It is simply a variable-focal-length lens that makes it possible to make a continuous change in the size of the object being photographed (zooming in to a larger view of the subject; zooming out to a view in which the object is seen smaller against its background).

The lens remains in focus at either extreme of the zoom, in or out. (A zoom lens should be focused at the tight end of the zoom — the largest focal length — and that will ensure that the subject will be in focus at all points down to the smallest focal length.) A typical zoom lens may be referred to as a ten-to-one (10:1) lens, which simply indicates that it is continuously variable over the range between a low millimeter number to a value ten times as much. Most small-studio cameras can be purchased with standard zoom lenses providing ratios of 4:1 or 5:1.

The major advantage of the zoom in small-studio production is that it makes possible the illusion of dollying into (or away from) a subject while the camera, in fact, remains in a stationary position. This is a benefit because lightweight camera mounts are almost impossible to use on a moving shot. Since they lack the weight of a broadcast TV dolly, there is a good deal of objectionable jiggling of the picture as the mount moves across the floor.

Shots are described in terms of their degree of closeness to the subject. Each director and cameraman will have a personal interpretation of a particular type of shot, but definition and agreement will soon come after working with each other.

The following terms are the basic ones used to indicate to the cameraman the type of shot wanted by the director.

Long shot (LS), a very wide shot which would, in the case of small-studio production, show the entire set. The long shot is used to establish locations, and to aid the viewer in knowing where he is in terms of the story and the action.

Medium shot (MS), an intermediate shot between a close-up and a long shot. In the context of small-studio work, a medium shot might show a person sitting at a table or a desk and include all of the person.

Close-up (CU), a shot in which the subject fills the greater part of the screen. A close-up of a performer would normally include the head and shoulders.

Extreme close-up (ECU), a shot in which the camera looks at a detail of the subject being shown. In the case of a performer's face, an extreme close-up might include only the person's eyes. The extreme close-up is very valuable in recording very small actions or details.

In motion picture terminology there are added refinements to this list of terms. They include extreme long shots (and refer mainly to exterior shots), medium close (somewhere between a medium shot and a close-up), and a medium long shot (half way between a long shot and a medium shot).

In addition to the basic terms, some directors will describe a shot of a person by where the bottom of the picture is to rest, i.e., a waist shot, a bust shot (or pocket shot). The director will also describe the shot by the numbers of persons to be in the picture: a *one shot*, a *two shot*, a *three shot*, a *group shot.*

Thus, a director may go from a two-shot to a one-shot by saying, "Give me a one-shot of the man on the left" or "Give me a medium shot of the man on the left." Both phrases mean approximately the same thing.

When dealing with performers, it is implied that the top of the shot should be slightly above the top of the person's head. Even this, however, varies with personal preferences. The space between the top of the head and the top of the picture frame is called "head room."

In most cases, the director and cameraman should add an additional 5% of space on both the top and bottom of the picture over what they finally want to appear on the screen.

There are two reasons for this. The first is that the television screen upon which the program will ultimately be displayed usually shows less than what is recorded on the tape. Secondly, in the transfer from a 1-inch master to a 1/2-inch duplicate, there is sometimes a slight loss of picture from top to bottom. Therefore, a director who calls for more "head room" is usually concerned about something being inadvertently chopped off the top or bottom of the picture.

There are two terms which refer to position of the camera in terms of its angle toward the subject. In most studio shooting the camera is mounted on a tripod or other mount and located at about the eye level of the operator.

Where possible, it is desirable to change this angle for the sake of variety and to enhance a particular mood or feeling that has been established in the production. There are also occasions when an angle shot will aid a great deal in the understanding of what is being attempted in the programming. The two main classifications are as follows:

High-angle. This means that the subject or the center of interest in the picture is *below* camera level. In other words, the camera is *elevated.* In regular broadcast television programming this is accomplished through the use of a boom to carry the heavy camera and the operator to a position above the subject or the action. With small-studio equipment a boom isn't required to raise the camera to a desired height. It can easily be lifted for the purposes of a shot or two (in this case on some type of mount on the top of a ladder, let us say), or it can be mounted on the type of tripod that can be elevated to a considerable height, i.e., about 8 or 10 feet.

Low-angle. This describes a shot in which the key subject or center of interest is *above* the camera level. In other words, the camera is at some height *lower* than eye level, and is looking up at the subject. This has more limited usage, but can be extremely effective as a change of pace in a production.

Another aspect of camera work is the condition in which the camera is moving, or appears to be moving, with respect to the subject or the center of interest in the screen.

These effects include the *dolly*, in which the camera physically moves toward, or away from, the subject. Broadcast television stations have cameras mounted on heavy camera supports that are designed to move very smoothly. As mentioned earlier, most small-studio tripods do not move smoothly and are apt to cause jerking movement, which is very distracting on the screen.

Another camera movement is called the *truck* or trucking shot in which the camera, on its mount, moves laterally to the subject — in other words, moves *past* the subject from left to right or vice versa.

A variation of this is the *following shot*, one in which the camera moves *with* a *moving* performer or subject. In small studios it is practically impossible to achieve this effect because there is too little space in which to maneuver the camera.

The effect of dollying in to or out from the subject can be achieved by the zoom lens or varifocal lenses, in which the camera *appears* to get closer to, or further from, the subject. The zoom can usually be handled far more smoothly than a dolly and should therefore be used. (One of the first bad habits that a cameraman has to overcome, by the way, is the overuse of the zoom lens, with every shot tending to have at least one zoom in or zoom out. Familiarity with the equipment, and considerable experience with the camera and viewfinder are the only remedies for this affliction.)

Another way that movement is imparted to the basic camera shot is through movement of the camera upon its stationary tripod.

Pan. The pan is the movement in which the camera is moved on its horizontal axis, scanning from right to left or left to right. In the small-studio situation, for example, a pan might be used to move the viewer from one speaker to another, to follow a man who walks to another point on the set, or to travel from the speaker to the object or illustration being discussed.

Tilt. The tilt is the movement of the camera in the vertical direction, scanning from top to bottom or from bottom to top. For example, the operator would tilt the camera up if a speaker on camera stood up, and would, likewise, tilt the camera down when the speaker sat down.

There are also other terms (mostly derived from motion picture production techniques) that are sometimes used in TV production. One such shot (and one likely to be used in a small-studio situation) is the *over-the-shoulder* shot, in which the person in a shot is seen over the shoulder of another person on the set.

Another kind of shot is a *reaction shot*, which may be a close-up of a performer's face displaying understanding of (or emotional reaction to) something that has just been said or that has just happened. The reaction might also be expressed in the shrug of the shoulders or an action with the hands, etc. An *insert shot* can refer to an extreme close-up of an item that needs to be seen in greater detail to achieve understanding, such as the numbers on a chart, a date on a calendar, printing on a page, or any other close-up view that helps visually to clarify what is being said.

The subject of *composition* is a difficult one to treat. The difference between "good" composition and "bad" may be very difficult to explain to the beginning cameraman, but with experience, certain habits and practices begin to emerge. There are very few hard, fast rules that can be laid down, but the following are some ideas that the beginning camera operator might consider.

1. Try not to let a vertical element in the setting (or other part of the scene) divide your shot into two equal halves. There is nothing dynamic about two parts divided evenly, so move the camera one way or the other.

2. Don't use the zoom lens unless there is some good reason to do so. Zoom into a subject to better understand it. Zoom out for the same reason. Remember also that a subject should be established before a zoom starts (stay on the shot for a few seconds before the zoom action begins), and remain on the tight or wide shot for a few seconds after the zoom is completed.

3. Don't continually frame objects from a straight-on position. A slight angle view is always more interesting.

4. Be sure to provide enough head room when shooting close-ups of people. Remember the masking effect of the receiver and its tendency to clip off edges.

5. When shooting someone who is moving about on a set, remember that movements toward the camera and away from it are more interesting visually than movements sideways to it (in which the person or objects always remain the same relative size).

6. The illusion of depth is better achieved if horizontal and vertical planes can be included in the shot (in contrast to working the subject against a stark, flat background). Such lines are inherent in the set, or the props, and include the angles of chairs or furniture, the edges of blackboards or curtains, or the presence of other performers.

7. Try for asymmetrical compositions. Perfect symmetry is boring.

8. If a performer is looking to one side of the screen, allow more framing on that side of the picture (that is, on the side where the camera would logically be if it were about to pan in the direction of the performer's attention).

9. Be careful about how the camera "cuts" the performer in the shot. Avoid uncomfortable-looking "amputations" of legs and arms.

10. Don't move on a shot unless there is some motivation for doing so. This includes dollies, pans, tilts, etc.

Following are basic working habits for the small-studio cameraman to follow.

1. Before the shooting session begins, make sure that the camera is securely fastened to the mount, and the mount to the tripod. Be

certain that the tripod is sufficiently heavy to support the camera or that the tripod is properly counterbalanced.

2. Look before you move. Make sure you won't run over a camera cable or run into something. The noise from such an accident can ruin the scene.

3. Put on your earphones and make sure they are operating. Communicate with the director or the engineer.

4. With the engineer's okay, uncap the camera.

5. Check to see that the head is in good working order. Adjust the head by unlocking it and then adjust the friction lock so that a certain amount of drag is evident when you tilt or pan the camera.

6. During production, wait for the director to release you from the shot. Also check the tally light and never move when it is on.

7. If you are operating a turret camera, move the turret as quietly and smoothly as possible. Racking the lens turret carelessly will result in unnecessary noise which may be picked up on the microphones, and may even cause damage to the mechanism.

8. Between takes, don't allow the camera to remain fixed on a bright area or pattern on the set. It could tend to cause the picture to burn in on the picture tube. Slowly move the camera about, if you are going to be off a shot for a considerable period of time.

9. To remove a burn, defocus the camera lens and leave it pointing at a medium-gray surface.

10. Be alert for new shots when the other camera is on the shot. If the production is very loose, the director will appreciate any imaginative work you can provide. If things have been rehearsed, then of course don't try to outwit the director.

11. Stay out of the field of view of the other camera(s).

12. If you're operating a dolly with wheels on it, make sure that you preset the direction of the wheels before you begin a moving shot. Otherwise, the wheels will be in the wrong direction, and the camera will jiggle considerably at the start of the movement.

Good camera technique was never achieved by accident, nor were first efforts imaginative or without flaws. But the results of practice and experience are readily noticeable, and contribute greatly to an efficient, creative production.

CHAPTER 10

DESIGNING VISUAL AIDS

One of the problems that must be faced by would-be producers of television programming is the need to support their efforts with professionally produced visual materials. This capability must either exist in-house, or provisions must be made to use the services of capable outside vendors. Visuals are too important a contribution to the success of ordinary programming to be overlooked or minimized.

The expert use of titles, charts, flip charts, photographs, diagrams, simple animation, and other illustrations can often make the difference between a boring recitation of facts, and an interesting, imaginative production.

Proper use of visual aids can make complicated information much easier to understand, and perform a real service to the viewer. This is a difficult point to keep in mind for many producers. They often have the idea that since they are already dealing in a visual medium, there is no need for additional pictorial information. This is a mistake.

There is also a mistake that can be made in the opposite direction, and that is using too much of a visual product that has been produced for other purposes. One of the common errors that people make is their attempt to get double duty from the illustrations used in a report, proposal, or some other written document. These types of illustrations are generally too complex for translation onto a projection screen or a television tube. What works well on the printed page is far too detailed for the purposes of TV.

The point to remember is that TV graphics are seen only once and then only for a brief period of time. There is no way for the viewer to "go back" and review the visual information. It is presented, and then it is gone. The idea, therefore, must be grasped immediately and understood completely.

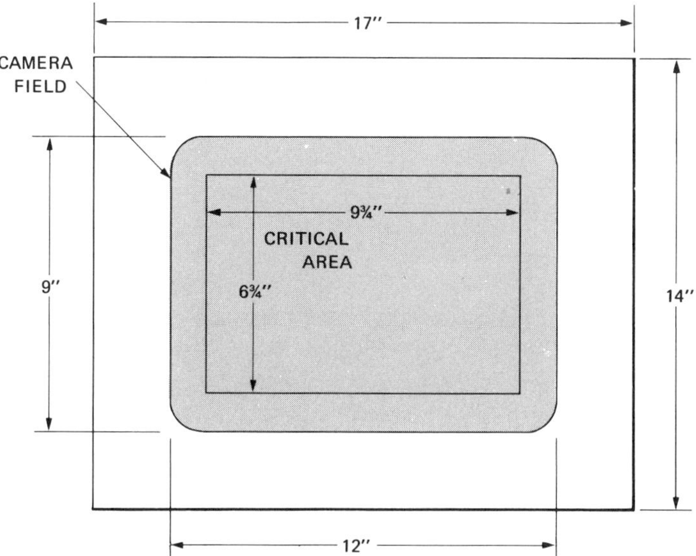

Illustration of working space and critical area for production of television graphic materials

The best advice is to *simplify* ideas, and use only the key ingredients of the intended message. This is hard work, and may require a good deal of negotiating with the authors of proposed visual aids. (All too often, a visual is merely an elaborate crutch devised by essentially lazy communicators.) The visual may be crowded with the key elements of what the speaker wants to say — with the frequent result that it bewilders the audience because it is so confusing and complicated.

Another reason for simplification is a purely mechanical one. The screen is of limited size. The small screen may be viewed by a number of people at one time, a few of whom may be seated in a disadvantageous viewing position.

A third reason is the nature of the television system itself. Small details are dropped out in the process of creating the image on the screen.

Following is a discussion of some of the key considerations to be made when planning television graphics. The subjects include: (1) picture ratio; (2) system limitations (in black-and-white or color); (3) the major types of visuals; and (4) basic lettering systems.

Picture Ratio. The ratio of television picture units is 3:4, — there are three units in height to four units in width. The total area of the picture is called the camera field, and this is the absolute size of the televised image. It must be remembered, however, that the picture viewed on the monitor is smaller than the camera field. This is because of the masking in the set itself. There are also conditions in which the set may be somewhat out of adjustment. At any rate, in the production of visuals for television, all graphic materials must appear in what is called the *critical* area.

In working with title cards, the following dimensions can be employed to determine the critical area. To begin with, the actual working space on the card is well within the outside edges. The overall size should be 14 × 17 inches. The working space should be 9 × 12 inches (the exact 3:4 ratio) and centered on the card. This is the camera field (also called the scanning area). Graphic content should be held in an area that is 6¾ × 9¾ inches. This is the critical area, in which none of the message will be lost by the masking of the TV set.

System Limitations. In black-and-white or color systems it should be kept in mind that TV has certain inherent limitations. In black-and-white, for example, the system can see, and render, only a limited number of shades of gray. Variations that are too subtle in the artwork for a visual will be lost.

Major Types of Visuals. The materials usually used in small-studio productions are title cards and super cards. Techniques with charts and maps, slides, crawls, and simple animation may also be used.

Title cards are produced according to the sizes and areas described earlier in this chapter. They include artwork such as logos, charts, graphs, quotations, key words, photographs, drawings, and other materials. They may be used to introduce subjects, as the titles for the beginning and end of shows, for identification of the producing group, and so forth. They should be characterized by simplicity of design, and by absolute economy of words used. They are designed to be shot by a studio camera.

Super cards are used chiefly to create the effect of white letters over a live TV scene, for instance, when it is useful to identify a speaker. The name and affiliation are supered (superimposed) on the bottom one-third of the picture. (It is a marginal note that in the preparation of super cards it should be suggested that the type used to show a name should be larger than that used for an affiliation.) A super card is prepared in the same size and format as a studio card, but the lettering or artwork appears only as white and only on a black background. When picked up by the camera, the black of the card will not reflect light, and only the white information will be seen. It is then electronically superimposed over the picture on the other camera. The production of super cards is the only time when such high contrast can be undertaken. In other artwork, the contrast cannot be as great. As with title cards, the least number of words should be used.

Charts and maps can be particularly helpful in explaining ideas on television. Often, however, they must be specially prepared to communicate effectively. Simplification is the main thought to keep in mind. A chart should have as little written information as possible, and should try to communicate a minimum number of ideas. For example, a chart showing five different lines, or trying to express five different values, is very difficult to understand, even if you are able to present each of the five ingredients in a different color.

By no means expect to use the sorts of charts and graphs that are normally used in written reports. They contain far too much title information for television, and often have vertical lettering along one axis. Prepare a special, simplified version for television.

The same is true of maps since most, as they appear in atlases or elsewhere, are far too detailed for easy understanding (and are protected by copyright). It will be necessary to redraw the map, showing only those key towns, boundaries, coastlines, etc., that are really of consequence to the speaker.

A *crawl* is a process in which the graphic material to be shown is prepared on a long, narrow strip of paper and placed on a large, circular drum. The camera is focused on a section of the strip and then the drum moves slowly, turning in a direction away from the camera. The effect is then one of a camera moving slowly down a long list of information. Motion picture and television credits are done in this way, usually on drums that are about 30 inches in diameter (with paper widths of about 12 inches).

A super crawl, of course, would be produced with white lettering on a black background.

A hand-operated crawl device will have a certain amount of flexibility as far as speed is concerned. A motor-driven crawl with a variable-speed motor will be much smoother.

Simple animation can be used to good effect in TV production. One of the simplest forms of animation is to prepare super cards, the message of which is slowly revealed to the camera by slowly pulling an all-black card across the front. Other techniques include placing a slot in the artwork, covering the slot with tape, and then slowly removing the tape from the back. A back light will then shine through the slot.

Photographs are quite easy to use as artwork. Most photographs, however, must be drymounted onto a stiff support so they can be used. This way they will not tend to curl under the heat of the lights. Also, if they are mounted flat there is less of a tendency to pick up unwanted reflections from the glossy emulsion. (If possible, they should have a matte finish.)

Slides can afford great flexibility in TV operations. If the facility you're working in has a film chain then it is much easier to put 35mm slides on the chain than it is to shoot studio cards. It takes less set-up time, the light on the object is not a problem, and the process frees a studio camera for other uses.

A slide should be produced in much the same manner as a studio card; that is, flat art can be shot from a standard-sized (14 × 17 inches) studio card. Hot-press titling can be used, as can any other system for producing lettering.

One of the advantages of slide production is that overlay cells (made of acetate) can be used without fear of getting reflections when the materials are shot. The lighting on the subject can be controlled far more easily. When studio cards are used in the studio situation there is overall lighting for the set, the participants, and for other subject matter. It is difficult to adjust lights specifically for the cards. When producing slides, a minimum number of lights (probably two) can be adjusted with great precision.

It is noteworthy to mention that when opaque letters are produced on a transparent material (such as acetate), then a whole range of materials can be placed underneath and used as background . . . different kinds of art papers, fabrics of various kinds, wallpapers, boards, bricks, sand, and any other material with an interesting color, pattern, or texture.

For black-and-white super slides, standard *black* on *white* original copy can be used. After it is shot, the *negative* can be used by itself (since it will then have the requisite white lettering on a black background). When using negatives of this sort, it is well to shoot the original on a type of film that is of very high contrast. These materials, called litho films, see only black-and-white and yield no intermediate gray tones whatever. Eastman Kodak's Kodalith film is a material of this type.

This technique can also be used for the production of materials in colors as well as white (or clear). This is done by the simple process of placing colored gels or cellophane over selected portions of the negative, and then binding the whole in a glass slide mount. They project extremely well on a slide/film chain, but care must be taken in selecting the colors used.

There is an alternative to using a slide/film chain, and it should be discussed briefly.

It is possible to use slides as part of the studio production. This is accomplished by constructing a small rear-screen projection system. This consists of a translucent screen upon which a slide is projected from the back. The screen can be any convenient size, and can be constructed in such a way that it becomes part of the studio set. A wide-angle lens is used on the 35mm projector to minimize the distance between projector and screen. A slide projected on such a screen will be bright enough to appear as a large piece of artwork (as long as the camera is close to the same axis as the beam of light from the projector; if the camera is placed at too great an angle to the screen, the picture intensity is diminished).

If slide projection is done in the studio, it is possible for a participant to operate the control that advances the slides, thus making cueing a good deal simpler for the director.

A word of caution: the noise of the slide projector fan and slide changing mechanism can be picked up by studio microphones. It will be necessary to cover the equipment with a sound-deadening device (called a "blimp"), and then be certain it does not generate too much heat.

Gobos offer an unusual visual effect in the studio, and can be easily made. A *gobo* is essentially a piece of artwork with a cutout in it, the camera viewing both it (the artwork) and the scene that can be seen through the cutout portion. A zoom-in will take the scene entirely through the "window" in the gobo, until the artwork can no longer be seen. (See example on next page.)

Hot-press titling is one of the simplest and most effective systems of title production. It consists, in the simplest terms, of a system for setting a message in metal type. The type is heated and then pressed against the board or other material to be printed on. Between the type and the board is placed a sheet of plastic foil which contains heat-sensitive ink. The heated metal letters transfer the ink from the foil onto the card in just a few seconds (and with the application of a certain amount of pressure).

A number of colors can be used in this process and printing can be done on acetate as well as opaque materials such as illustration board, etc. The other advantage of the system is that it is a dry process and there is no down-time required for ink drying.

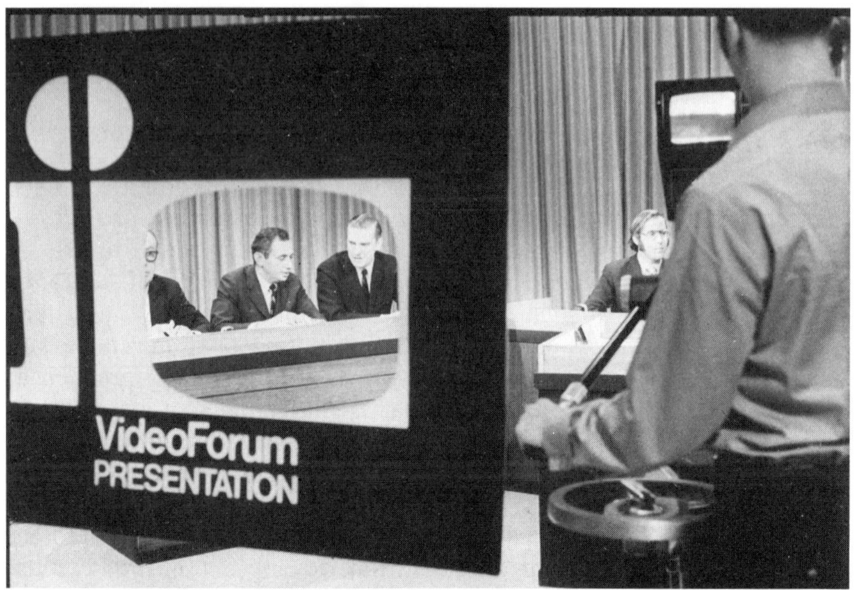

(Photograph courtesy Institutional Investor Video, Inc.)

Gobo, in which the opening shot is framed by a cutout and artwork

Handlettering is a method of production, but unless the person attempting to handletter a card or other visual is really proficient and expert, the process can take inordinate amounts of time (often with less than satisfactory results). Nothing looks worse than poorly executed handlettering.

Rub-on letters can be used easily and, with some practice, with good results. They are printed on a thin plastic sheet, and when the letter is burnished onto the artwork surface it is transferred. Care must be taken with this system since the letter can't be moved once it is placed down. (It can, however, be removed entirely by applying cellophane tape to the letter and lifting it off the artwork surface.)

Lettering guides offer a means of producing letters, but it generally requires considerable skill and experience to use them effectively. One of the best systems of this type is the Varigraph, and it works on the pantographic principle.

Printing processes employ regular lead type which is set (face up in reverse), locked in place, and inked. The artwork surface is then placed on the inked type and a roller is run over it, thus transferring the inked image. *Linoscribe* is a small, and comparatively inexpensive, system of this type.

Plastic letters represent a simple system that employs plastic or metal letters which can be arranged either on a flat surface or placed in a slotted feltboard surface (so that the final effect can be shown vertically).

We have made no strong recommendations in this chapter concerning gray-scale comparisons, or the hues or intensities of colors to be used in color television systems. If only one piece of advice could be given on this score — advice to those who will design materials for TV use — it would be the same as that for TV camera operators: use a particular system and discover its capabilities.

There is very little to be learned from books about the process of developing successful materials or techniques for programming — and nothing at all to be learned from talking about them. The TV cameraman must *use* the camera to learn its capabilities and limitations. Likewise, the designer must spend time in the studio — often with paint and brush in hand — to *see* on a monitor the effect of certain styles, kinds of type, combinations of colors, and basic techniques. He or she must test the materials on the spot, not guess about them as they are tacked on the drawing board.

The TV graphics designer has one notable advantage that film graphics people don't have. Since there is no film or laboratory expense involved in checking a piece of artwork, tests can be made at no cost. Also, there is no delay in seeing the end result. He or she can see the effect of the work immediately, and make any necessary change in terms of what is seen.

CHAPTER 11

SETS AND PROPS

Sets and props for the small studio are an important element, but also one in which simplicity must be the key. There isn't the necessary space to construct, erect, or (later) store settings that are very complicated, massive, or detailed.

The purpose of the TV studio set is to accommodate the needs of the performers and the story they are trying to tell. The setting should be uncluttered so that the cameras can easily work in it and around it. The setting should also be arranged in such a way that it is not difficult to light. Those are the needs of the setting when in actual use. There are also some off-camera requirements of a setting — that it be simple to disassemble, handle, and store. In small-studio situations there just isn't room or necessity for cumbersome sets, or setting areas that are crowded with furniture, set dressing, and props.

The *set* consists generally of the background of the scene being shot, and the furniture that may be required in front of that background — including chairs, tables, desks, lectern, and so forth. *Set dressing* consists of design elements on the background itself: drawings, paintings, charts, etc. — and such things as desk sets, water carafes, glasses, etc., on the set. *Props* (properties) include any of those things handled by a performer in the course of the production, and include such things as models, pointers, and hand-held pictures or charts.

The greatest amount of work in a small studio will probably be done against a draped background. This provides a simple and attractive setting, and one that is relatively trouble-free once it has been installed. The highlights and shadows of the pleats in the drapes add a depth to the camera shot that must be achieved by other means if you are shooting against a flat surface.

If drapes are installed, they should be of sufficient height on the set (approximately 10 feet), and it is desirable that they be mounted

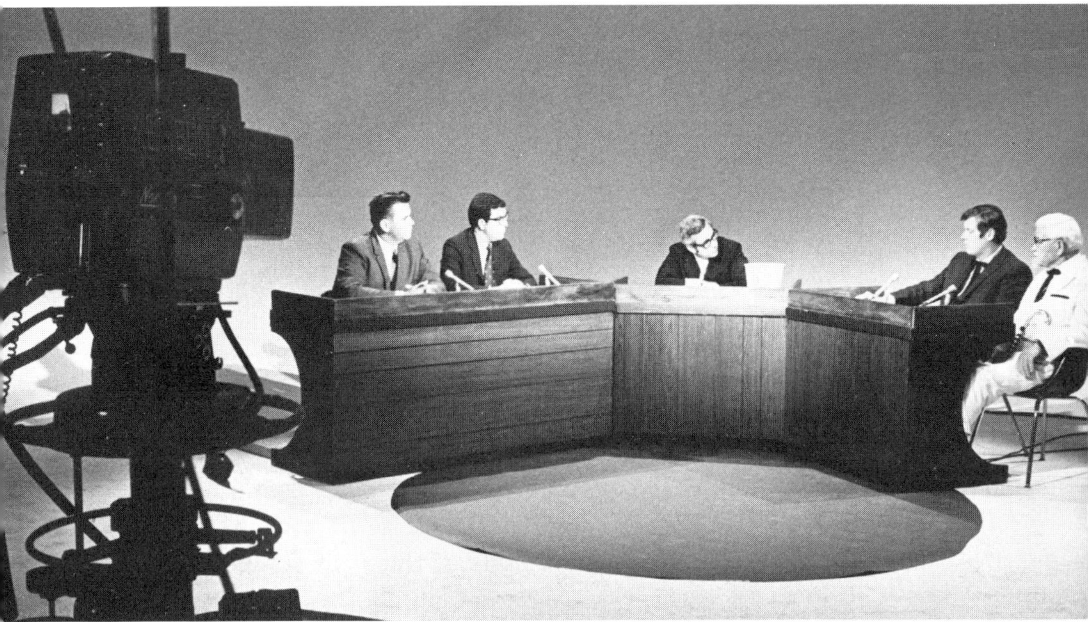

(Photograph courtesy Institutional Investor Video, Inc.)

Simple set in front of a scrim, also showing use of circular floor design to create visual interest

on tracks so they can be moved back and forth. It is also useful if the section of drapes in a corner can be made to curve gently rather than describe a sharp angle. The curving background will create a more pleasing picture, and will offer more flexibility for the placement of furniture and determining camera angles.

Drapes should carry well-defined folds or pleats to create a rich textural effect (and this won't be the case if they are stretched too flat). They should be lit from the side rather than straight on to enhance the drape effect. The drapery material should be of intermediate weave and be of a color that will reproduce as a rich, medium gray on black-and-white systems, and not clash with flesh tones in a color television system. (A medium blue is always a safe bet.)

An alternative to drapes as a background is the use of seamless paper, which is available from photographic and display houses. The paper is produced in numerous colors and is delivered in rolls that are 9 feet high, with widths up to 36 feet.

Seamless paper can be tacked or stapled directly to a wall, or it can be fastened to flats that have been slightly angled. The paper can be used as is, or it can be painted with water-based paints such as scene paint or tempera.

In some cases it will be advisable to build a setting for a special show or series of shows. The length of time you expect to use the set will determine the kinds of material from which it should be made. In other words, if it is to be a permanent and regularly used set, it should be made of a durable material that can withstand the use. If, on the other hand, it will only be used one or two times, then a much less durable, and far cheaper, material can be substituted.

A set can be constructed from a variety of materials that range from the extremely solid type (made of wood, plywood or Masonite) to the light, theatrical-type setting (made of a wooden frame covered with canvas). Settings can also be made from paper and cardboard.

No effort will be made in this chapter to outline the process of constructing sets of the kind used in major productions. These are units that consist basically of 1″ × 3″ framing which is covered with cotton canvas (or a heavier material such as Masonite). Information on the design and construction of sets of this type can be found in books dealing with theatrical design.

Rear-projection slides that create the entire background can also be used, but are difficult to adapt to a small-studio situation. If you contemplate using slide projections, it is best to contact a manufacturer who specializes in this type of equipment.

A number of design effects are possible with cardboard. Two types of board can be used — the regular material from which cardboard boxes are made, and a thicker, more durable version (a brand name of which is Triwall). It can be tacked to 1″ × 2″ lumber for support and can be painted with altex-type paints. Edges and corners can be taped with a material called duct tape (also called gaffer's tape or silver tape), which is a wide tape and one that takes paint very well.

Another background setting technique is the use of a *cyclorama* (cyc). This is a continuous piece of cloth (muslin, cotton, twill) that is stretched along a wall, in a straight line, or at times with a slightly curved effect. The cloth is usually an off-white color and it creates the idea that there is an infinity of space in the background, i.e., no back wall at all. To operate successfully, the furniture and other set materials must be placed a few feet in front of it, and the backlighting must be very even.

The feeling of infinity of space can also be achieved by the use of the bare studio and the selective placement of lights. If the light is restricted to the participants themselves, and the background remains dark, the effect is more dramatic than with the use of a cyclorama.

This technique, that of a dark, indistinct background, is called a *limbo* setting. It requires the talents of a good lighting director, or some creative experimentation, but is very effective if done well.

Whatever the setting chosen, it is important to avoid overlighting it. The background should usually be darker than the foreground so that the performers and subjects stand out against it.

When a fully lit set is used, the furniture on which the talent will sit should be placed at a distance of at least 6 feet from the background in order to create the necessary feeling of depth. The closer a chair is to a flat surface, the flatter will be the overall appearance of the entire picture when projected on the television screen. Another problem created by placing furniture and talent too near the background is that light bouncing off the background (particularly if the wall is colored) affects the flesh tones of the participants.

In most small-studio situations, actual furniture can be used. The only basic rule is that the furniture be of simple style, relatively light weight, and not call attention to itself.

A peculiarity of television is that it doesn't display relationships of people exactly as they are — the screen will show them farther apart from each other. To compensate for this it is necessary to place objects and people closer than they would ordinarily be to each other. For example, in a two-man discussion, the arms of the two chairs should be touching. This is another reason why bulky pieces of furniture should be avoided.

Not all set pieces need to be the real object or fully constructed — only those portions that will be seen by the camera. A drawerless table with a cardboard front can serve as a desk.

The first consideration you should give to dressing the set is that the objects you select should look natural in the setting and also not call attention to themselves. A picture on the wall should be fairly nondescript in subject matter, color, and design. A fairly safe bet are photographs of products and facilities of the company. It must be kept in mind that set dressing is part of the background. Therefore avoid such things as using the company logo on a lectern, or placing a name plate on a desk.

To create the feeling of a natural setting, certain things should be placed on the set to make it look as though it is being used — such desk items as ashtrays, pens, papers, and so forth. On a table next to a chair in which a performer is seated, can be placed an ashtray or carafe and water glasses. Potted plants also make interesting set pieces.

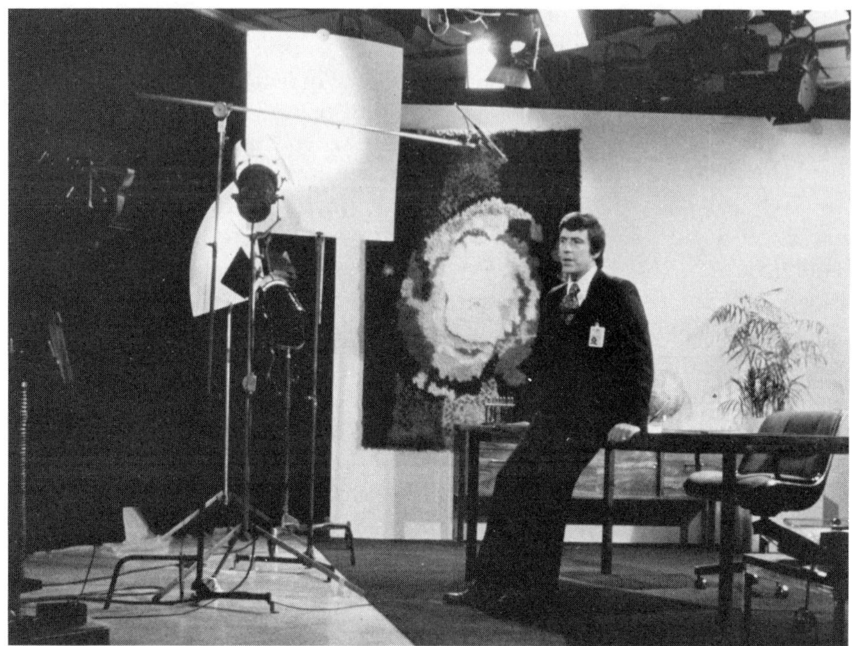

(Photograph courtesy International Business Machines Corp.)

Wall hangings, plants, and desk props help create a natural setting

Properties, handled by performers in the course of a scene, contribute to the believability of the action. Models, pointers, pencils, books, and photographs can be used effectively as hand props.

The difficulty with props is that their use must be carefully rehearsed. The director must know exactly how they are going to be handled, because they sometimes require the use of a close-up shot. In these instances, the performer must also be advised when the camera is on the prop so he or she can hold it still. (A pointer must be held at a precise place on a chart until someone signals the performer that the shot is completed.)

The use of photographs or hand-held charts by the performer must be rehearsed so that they are placed at the proper vertical angle to avoid light reflections, and in a position that can be easily and properly seen by the camera.

As with all other aspects of production, the successful use of settings and properties includes seeing them projected on the screen before the actual taping. They are correct if they achieve the effect you wish, while not detracting from the content of the show.

CHAPTER 12

LIGHTING TECHNIQUE

This chapter deals with three broad aspects of small-studio lighting technique: (1) the basic philosophy of illumination; (2) the various types of lighting effects required; and (3) the kinds of lighting equipment needed to achieve a particular effect or result.

Philosophy. The process of lighting a television set, or a portion of a television studio, is an altogether different process from illuminating a classroom or a workshop. In the first place, a different light level is required in order to operate the TV cameras effectively.

Light is measured by a light meter and the resulting level is indicated in footcandles. Classrooms and offices may be found to vary somewhere in a range of 10 to 75 footcandles. Studio lighting for vidicon cameras requires a light level of about 200 to 250 footcandles.

If insufficient light is used on the set, the effect is immediately noticeable on the television monitor or on the tape. The chief problems created include "noise," or graininess, and the inadequate rendition of blacks, whites, and gray scale. (The tendency in an underlighted set is to increase the gain of the camera. This doesn't solve the problem, and only more light can be of any help.)

A properly lighted set will have a pleasing range of grays all the way from black to white — which also provides better resolution and focus (because the camera lenses don't have to open up so wide), and results in a much crisper picture.

Many studio technicians have found direct relationships between TV studio lighting techniques and the techniques of still photography or film production. The lighting is the same in all major ways, and both the TV producer and the motion picture producer must concern themselves with the proper quantity of light, and also the *color temperature* of the light.

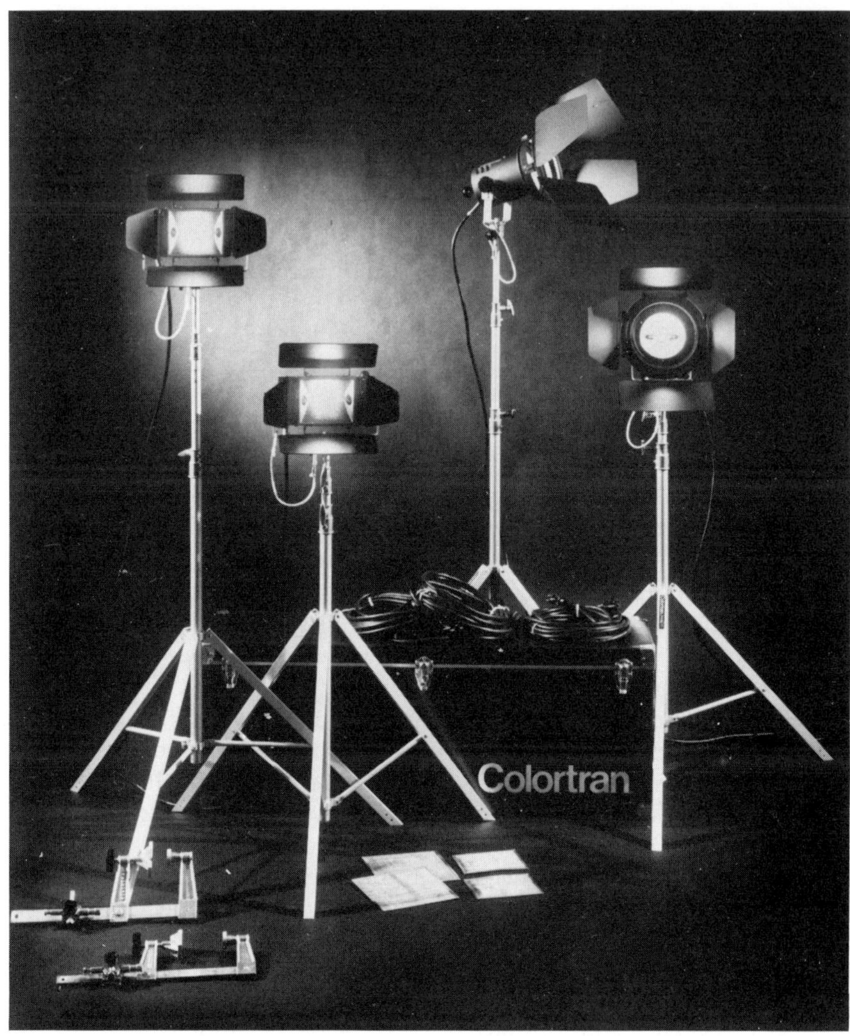

(Photograph courtesy Berkey Colortran)

Portable lighting kit designed to illuminate a 10 ft by 10 ft by 10 ft area to 150 footcandles

To properly expose color motion picture film, or to properly light a set for color television, it is essential that the light on the subject being shot be of a constant color temperature. Without going into detail, we can say that color temperature is a measure of the blueness or the redness of the light source. Regular incandescent lights are more reddish, for example, than fluorescent lights, which are quite blue. (Color films are balanced either for daylight or for artificial light, and the photographer chooses to use one or the other. If daylight falls through a window onto a scene that is being lighted with artificial light, then large color filters must be placed over the windows to balance the color temperature of the sunlight with that of the lighting equipment.)

There is no such problem of color temperature in black-and-white TV lighting. It is quite possible in a location situation to mix daylight, fluorescent light, and light from portable lighting packages. One must be careful and consistent, however, in the level and temperature of light for color TV.

In the studio it is a must to have absolute control over the quantity and quality of light. For example, there should be no windows in the studio through which daylight falls. The condition of the outside sky (cloudy, sunny, etc.) would make standardization of lighting an impossibility. In other words, no lighting set-up would be consistent from one day to the next. When one has finally achieved satisfactory light on a set that will be used frequently, one should be able to reproduce that effect with a minimum of effort. This is done through the use of a *lighting plan*, which will be explained later.

It is quite important to maintain sufficiently high levels of light so that the studio cameras can operate at least in the midrange of their f-stops.

The appropriate level can be easily measured by a photographic light meter (and the meter should be read at 1/30th of a second, since that is the equivalent speed of television systems).

Generally speaking, the eye is an excellent judge of the relative quality or proportions (but not necessarily *intensities*) of light that is falling on the set. If the eye sees dark areas, or harsh shadows, then you can be sure that the television system will see the same things — and perhaps even accentuate them. The use of a light meter during one's first days at trying to light a set will prove invaluable. By moving from place to place on the set it will be possible to get instantaneous readings of the level, and to correct discrepancies at the same time. Testing the effect with cameras is, of course, *the* determining factor.

When lighting the main working areas on a set, one should constantly try to achieve pleasing skin tones on the faces of the performers. Along with makeup, effective lighting is the only way to produce a natural-looking appearance. Strong shadows on the face should be avoided (in color TV, as a matter of fact, those shadows can take on different colors), and reflections from clothing and sets should be avoided. The use of gels (described later) should be avoided on the performers, and restricted to use only on the setting.

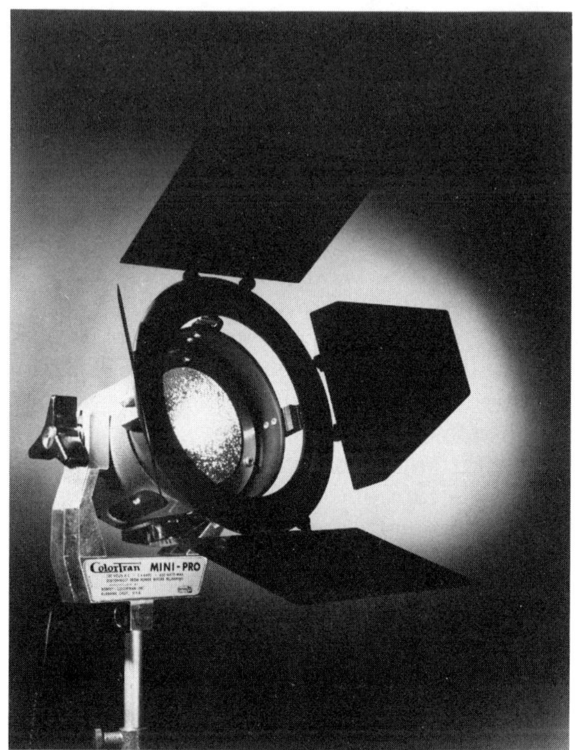

(Photograph courtesy Berkey Colortran)

Key light with both spot and flood focus

Types of Lighting Effects. The main purposes of studio lighting are to produce a natural appearance on the part of the performers, to provide sufficient light to operate the electronic systems, to limit the contrasts between the highlight areas and shadows, to achieve the appropriate mood for the programming in production, and to indicate form and dimensions correctly.

The *key light* is the principal source of illumination for the performer or subject on the set. It is usually positioned in front of the subject and above it. It is a directional light, such as an ellipsoidal spotlight, and is intended to produce shape and form in the performer's face. The positioning of this light is very important. If set too high there will be a tendency for the performer to have dark shadows in the eyes and appear to be very hollow-cheeked. Too low a setting creates a washed-out effect.

Fill light is the general illumination used to provide lighting in the scene. The largest number of lights in the studio are designed for this purpose and include flood lights and scoop lights of various types.

Back light is the illumination that falls from almost directly above the performer and serves to highlight the top of the hair and shoulders. By doing this it tends to separate the performer from the background of the set. If only key lights and fill lights are used, there is a tendency for the performer to blend into the rest of the scene and become very two-dimensional.

A *background* or *set light* is arranged specifically to light the walls of the set, or the background generally, and may consist of flat, even light, or light that projects a pleasing pattern. To avoid the monotony of large, evenly-lighted or evenly-colored walls, the use of the "cookie" (also called "kukaloris") is advised. The cookie is made from a flat piece of metal with cutout areas in it. When positioned in the front of a spotlight it projects the cutout pattern on the background.

A *camera* or *eye light* is a small cliplight, usually of about 150 watts, mounted on the camera. It improves the lighting on camera cards, and also adds sparkle to the eyes of the performers. (These lights should be turned off if the camera advances to a point very close to the performer as it will otherwise tend to produce a very washed-out appearance.)

Lighting Equipment. The two basic types of lighting instruments are floodlights and spotlights. Floods are used to produce even light for general illumination of the set and the background. They have no lenses, and there is no way to control the beam of such a light. The only measure of control over their output is the choice of the number of instruments used.

Floodlights are sometimes called scoops, because of the scoop-like appearance of the reflectors. Larger varieties have an 18-inch reflector and utilize a 1500- or 2000-watt bulb. Quartz scoops are considerably smaller and have a reflector with a diameter of about 12 inches. They use a 500-watt quartz lamp, and represent a better choice for a small-studio operation.

Spotlights are used to deliver light to specific areas on the set, and to illuminate key performers or subjects. They are made with and without reflectors, and have systems to adjust the spacing between the lamp and the lens in order to control the spread of the light beam.

Spotlights are further controlled by the use of barndoors, which work like moveable doors to crop the light beam; and snoots, which

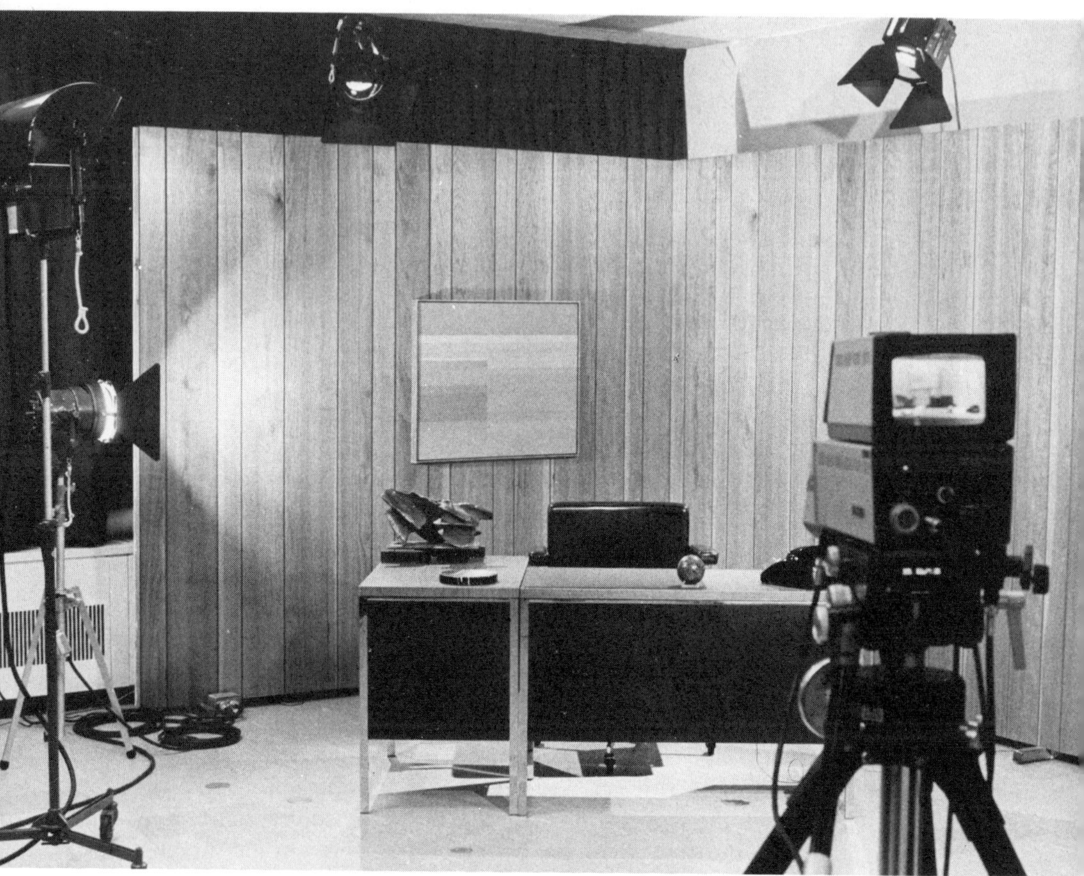

(Photograph courtesy Chase Manhattan Bank, N.A.)

Small-studio setting showing moveable lighting system

serve to further restrict or focus the beam of light falling on the subject.

Spotlights usually fall into one of the following classes: ellipsoidal or Fresnel. (A third class might be the internal reflector spotlight, which is actually a lightbulb with a reflector inside the bulb. It has no focusing capability, but can serve a fairly useful — if limited — function in general lighting.)

An ellipsoidal spotlight is one in which the light from a powerful lamp is reflected by an ellipsoidal reflector and projected through a lens. Between the light source and the lens are metal shutters which can be used to control the shape of the light beam with considerable

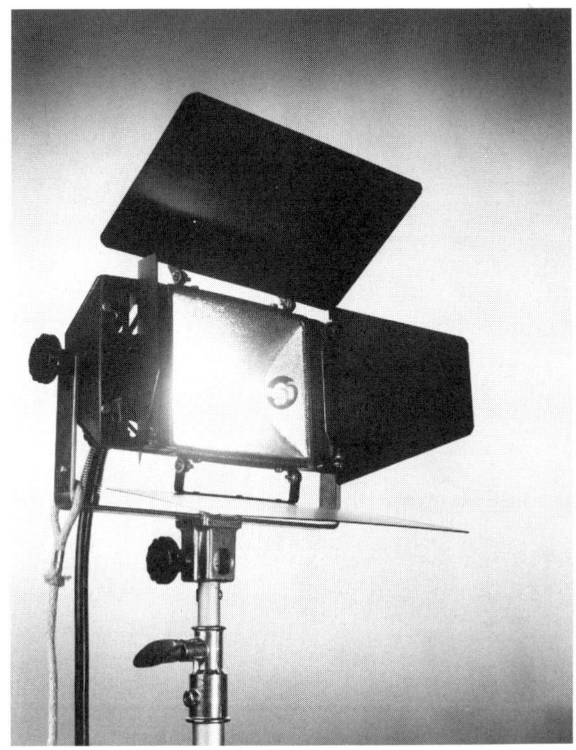

(Photograph courtesy Berkey Colortran)

Fill light with barn doors

precision. Regular incandescent types of these lights range in size from 500 to 3000 watts. Quartz versions, with beams which are not quite as sharp or as easily controlled, are available in sizes from 400 to 2000 watts. A spotlight of this type should be used when projecting "cookies" on the back of a set.

The Fresnel spotlight is so called because it utilizes a Fresnel lens. The bulb and reflector unit of this spotlight are mounted on a track. A focusing handle moves the bulb and reflector closer to (or farther from) the lens, thus adjusting the light beam anywhere from a flood effect to a sharp spot. Spotlights of this type are available in incandescent models in sizes from 150 to 5000 watts. Quartz versions are made from 400 to 2000 watts.

Lighting a set can be a time-consuming proposition, mostly because so many lighting instruments will be used at one time. If a show is to be done on a continuing basis it is a good idea to prepare a lighting plan, which is nothing more than a precise record or chart of what lights are used, in what position, and at what intensity.

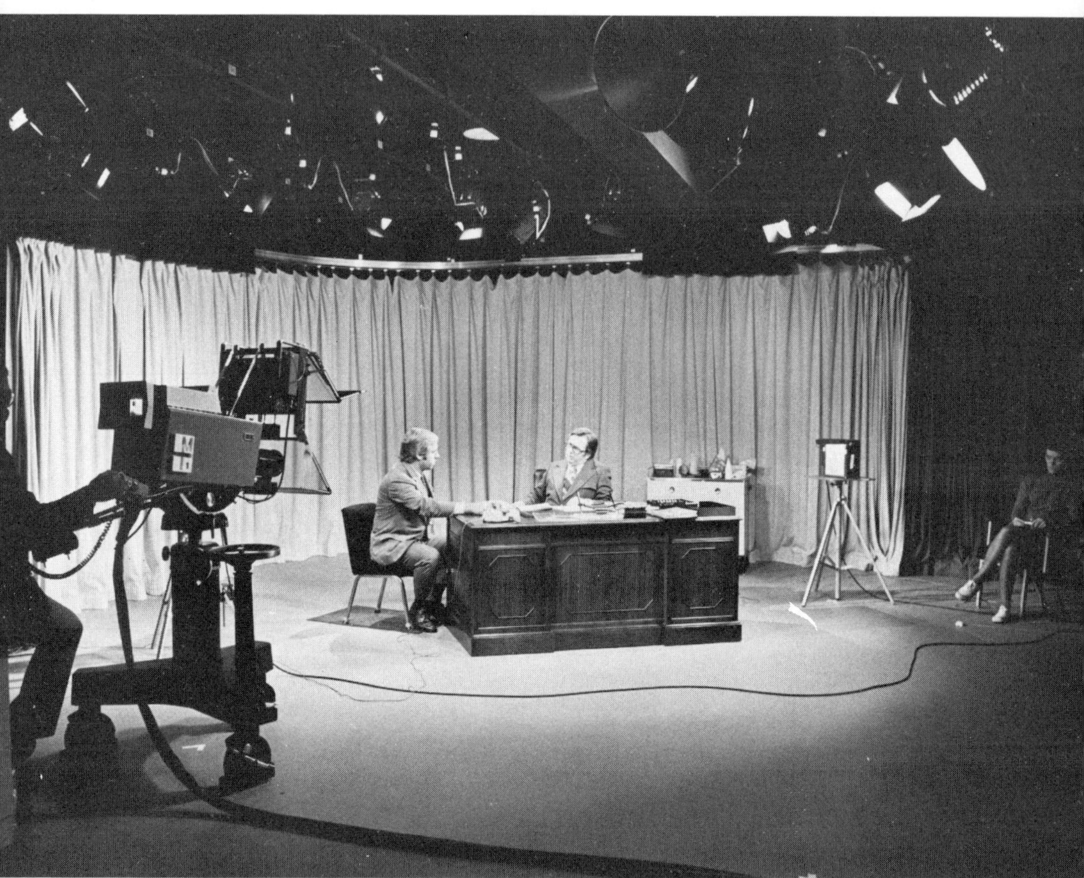

(Photograph courtesy Merrill Lynch, Pierce, Fenner & Smith, Inc.)

Studio shot showing overhead light racks

Fluorescent Lights. It is our experience that fluorescent lights should not be used in the small-studio situation. Their output is very hard to control, they are apt to create electronic disturbances (hum and flicker), and their color temperature is such that they can't be used for color television.

Gels. Gels (gelatins) are the transparent coloring medium for lighting instruments. They were developed for theatrical use and are produced in a large range of colors. They can be used in color television production, but only in areas outside those in which performers appear.

Mounts and Supports. One of the most common methods of installing lights involves the use of a gridwork of pipes. The pipe, about 1½ to 2 inches in diameter, is arranged in parallel lines or mounted crosswise, and the lighting units are mounted either directly on the pipes or on hanging brackets that connect to the pipes. In this technique the entire electrical power distribution system is above the floor and out of the way of cameras, sets, and performers.

The most common mounting device is a "C" clamp. Certain lights that will be used in a specialized way (such as key lights) can be mounted on a pantograph, which enables the light to be easily moved up and down.

Patchboards. When the design of a small studio is being developed, it would be wise to consider the use of a patchboard to simplify the operation of various lights, dimmers, and switches. A patchboard is nothing more than a switchboard-like assembly of connections for lights and switches (or dimmers). Short patchcords are used to connect one to the other and complete lighting circuits.

Dimmers. You may consider the use of dimmers when the electrical distribution system is in the design stage. They will make possible a considerable amount of flexibility in the control of the studio light. It must be remembered, however, that changing the intensity of the light source by dimming it also changes the color temperature of the light. This will cause problems in color television, but none in black-and-white. Color television lighting won't tolerate the use of dimmers on any of the lights used for the talent — because it would radically change the skin tones. The use of dimmers is restricted to lights that fall on the set, or on objects other than the performers. (To lessen the intensity of a light falling on a performer in a color television set-up, a stainless steel scrim (screen mesh) can be inserted in front of the lighting unit. This will cut down intensity without changing color temperature.)

CHAPTER 13

THE PRODUCTION PROCESS

This chapter is a discussion of the step-by-step actions you must undertake for the actual shooting of a television production. Although the discussion assumes a one-day shooting schedule in an in-house studio, certain steps (scenery and lighting, for example) can be accomplished a day or two before the day of production if scheduling permits.

Another assumption we have made is that you have preplanned your production so well that everything concerned with the show is ready and at the studio. The sets are constructed and painted; the visuals are completed; the performers have been notified when to arrive; there are spare bulbs for projection equipment; a stand-by cameraman is available.

The following is a simple check-list of production steps that must be taken before, during, and after the actual shooting of a television show. A detailed description of each step follows this listing.

1. Clean the studio and control room.
2. Erect scenery in studio.
   - Clean VTR and check out control room equipment.
3. Light scenery.
   - Add furniture and props.
   - Give visuals to film-chain operator (or determine studio set-up).
4. Check out cameras and control console — set video levels and balance cameras.
   - Place microphones and set approximate levels for all audio.
5. Determine camera positions.
   - Check scenery in relation to camera angles.

- Shift scenery furniture or props, if necessary.
- Shift lighting, if necessary.

6. Rehearse camera shots including dollying and trucking.
   - Secure scenery.
   - Mark furniture position.

7. Rehearse film and slide chains (or studio projections).

8. Meet with performers to discuss show.
   - Production crew takes a break.

9. Walk through movements with performers.
   - Confirm sequence of visual aids.

10. Rehearse opening and closing camera shots.
    - Rehearse use of visual aids.
    - Record sample shots and play back to confirm VTR is operating without interference.
    - Make up performers.

11. Bring in talent to rehearse specific sequences.
    - Clear studio of nonoperating personnel.

12. Check all audio levels.
    - Rehearse all audio cues (music, narrator, etc.).
    - Check makeup for final powdering.

13. Tape the show.

14. Shoot cover shots.

15. Review tape in conference room.
    - Crew stands by.
    - Kill studio lights.
    - Cap lenses of cameras.

16. Reshoot required sequences.

17. Thank crew and talent.

18. Cover equipment and put in storage rooms.
    - Store properties (and scenery, if necessary).

19. Clean the studio and control room.

At various times throughout this book we have emphasized the importance of maintaining a clean studio. The primary reason for our concern is to save you money. Dust particles in the equipment are one of the major reasons for equipment failure or "noisy" pictures. A speck of dirt that interferes with the transmitting of an electronic signal on a 1-inch video tape is proportionately magnified on a 19-inch television screen. In addition, a clean, waxed floor means easier handling of camera dollies and cables. After the studio has been cleaned you can then remove the covers from the cameras. In the control room, the cover can be removed from the VTR. It is then cleaned, with special emphasis given to removing oxide particles from the video heads. This cleaning procedure should be performed before every taping, even if more than one show is being recorded during a given day.

Once the VTR is cleaned and ready, the other control room equipment and console can be turned on and checked out.

A simultaneous step with the control room check-out is the placement of the scenery. Naturally, if the sequence is an on-the-spot recording at the location of an event, then the scenery is already determined and the positioning of the cameras is the major consideration. However, we assume most recordings will be performed in a studio, and will require some type of background or scenic design. Chapter 11 (Sets and Props) has already discussed the preparation of scenery, but it now must be put in place before the crew can proceed to the next step.

The scenery should first be placed only in approximate position, with the idea that it may need minor shifting later because of camera angles or to make extra room for furniture. This also applies to lighting the studio, since a performer's physical characteristics (such as a high forehead) may require minor lighting changes.

Lighting a studio or set for a first time can be a lengthy process, and ample time should be allowed. If you are still lighting the scene when the performers are assembling, you run the risk of extended delays, overheating of the studio, retouching makeup, and concern on the part of nonproduction people who wonder why it takes so long to get started. Adequate time for lighting should be planned so that the lights can be completely turned off and the air in the studio cooled just prior to shooting the show.

When setting the lights, it is helpful to have someone sit in for the performers and go through any anticipated movements. In this manner, the lighting director can see that the light falls evenly on the entire setting, that "hot spots" do not occur, that the lights are not in the performers' eyes, and that noticeable shadows are eliminated.

Once the lighting has been determined, the camera levels in response to the lighting can be determined and adjusted accordingly. This is performed jointly by the cameramen in the studio and the technical director in the control booth.

When two or more cameras are used, they must be "balanced" — that is, they must have the same target level, the same gain level, and be generating an equal video level. This is to assure that when switching between cameras you are recording the same level of brightness and contrast. Cameras should be balanced using the video waveform monitor as explained in Chapter 4 (Equipment). If a video waveform monitor is not used, then the cameras must be balanced by projecting their pictures on a control room television monitor and making adjustments from what you observe. Adjust your cameras to achieve sharp resolution of the image and strong contrasting colors (or shades of black and white). Because of losses from reproduction and the vagaries of playback units, the final projected images will ultimately be slightly softer and the colors more muted. The stronger they are at origination, the less the loss will be noticed.

At this point the cameramen "play" with the cameras to get the feel of the friction locks on the tripod head and adjust them for smooth panning and tilting. A certain amount of resistance is desirable so the movement is not too fast or uneven. The cameraman also gets the feel of the lens changes and how much area each lens position encompasses.

Meanwhile, the microphones can be put in place and approximate audio levels can be set. It is assumed that different persons speak with different itensities and final adjustments cannot be made until the actual performer is in position at the microphone.

No matter how well planned camera positions may look on paper, they must be practiced and resultant pictures seen on the screen. There are many factors of scenery and lighting that cannot be predicted until they are actually viewed by the camera. For example, a picture on a wall may be directly in line with one of the camera angles on a performer so that the picture appears to be attached to the performer's head. Either the picture, the performer, or the camera must change position. In another case, the seam of two pieces of scenery that join together might be directly behind the performer from a particular camera angle. A simple camera move to a position one foot in either direction might remedy the situation.

It must now be apparent that a stand-in for the performers is also helpful at this point in the production. The stand-in can help set microphone levels, and the director will be able to spot small distractions in scenery and properties.

This is the time for the adjustments in scenery, furniture, properties, and lights. And this is the time period that goes most slowly, because a shift in any one of these elements means adjustments in the other. But it is these seemingly insignificant refinements that eliminate the amateurish look of a small-studio production.

Once the scenery and lighting are established, they should be made secure and their positions marked.

The flats or drapes should be sand-bagged.

Chalk-marks or tape on the floor will show the positions of furniture legs in case they should get moved. Similar marks may be used to show the positioning of a set that may be taken down and put back later.

Lighting clamps should be checked, and screws tightened.

A light plot should be drawn and rheostat settings should be noted so that future shows can benefit from the lighting experience of previous ones.

Again using stand-ins for the performers, the director is now ready to rehearse the series of shots that have been anticipated. This step is especially critical if there are changes in camera position during the running of the show. The camera operators must check to be certain that they can, in fact, make the move in the permitted time; that there are no obstructions in the way of the move; that the cable lengths are sufficient and cables are not crossed; and that the stage crew is in position to assist in changing cards, etc.

Once the studio crew is set and rehearsed, attention can be turned to the film and slide chains. The film must be ready to go and there must be agreement on how much warning (either in seconds or leader frames) the film chain operator will be given. These cues to the operator must be rehearsed and he or she must actually start the film to be certain the equipment is projecting and recording correctly.

Any slides to be used must be in a prearranged order, and these also must be checked and positioned on the screen.

If studio cards are being used instead of a film or slide chain, these should be rehearsed at the time the camera shots are being set and checked. One member of the stage crew must be assigned to handle the cards so he can get a sense of order and timing.

While the camera and stage crews are taking a break, the producer and director can meet with the performers in the conference room. The discussion will first of all center around the rehearsal steps necessary before the show is actually taped, so the performers can get a sense of participation.

The sequence of actions and movements is discussed along with the sequence of visual aids. If the show is merely a discussion, then the order of topics and speakers should be determined. Even an ad-lib or extemporaneous show needs a predetermined beginning and ending, even if the middle is completely free-wheeling.

First-time performers should be cautioned about being distracted by the movement of cameras or personnel. Their attention must stay with what they are doing or the conversation taking place around them. They should be told when to look at the camera (when being introduced) and when to ignore it (most other times).

This conference room discussion can be most critical in terms of putting performers at ease. Therefore, the conference room is not the place to go over any technical problems the crew may be having in the studio. The discussion should center around what has already been decided in the preproduction conferences and any adjustments that may have to be made. The producer or director must assure the performers that everything is under control and that this review is merely to save time in the studio.

If there are no problems or disagreements that need to be resolved with the performers, they can be taken into the studio so they can view the area in which they will be working. The production crew can be on break during this period, so the studio will be reasonably quiet. Only the work lights will be on and the performers are free to wander around and become familiar with the positions of the furniture, the props, and the cameras. They can check the visual aids in the studio and the sequence of charts or slides.

If everything is all right, the performers can leave to receive makeup, and the camera and stage crew can return for further rehearsal.

Elements to be rehearsed can be the opening and closing shots of the show, and practice with the cards or slides that give the credits. The opening music and narration can also be rehearsed so that cues and timing are well established.

During this period of rehearsal it is also advisable that the VTR operator record the opening and closing sequences, even if they are later to be recorded "live." This permits the operator to play back the tape to be certain that the VTR is recording properly and that the picture is stable and the resolution is sharp.

By this time the performers should be ready to enter the studio to prepare for the actual taping. It is even sometimes advisable to let them watch some of the preliminary tapings so they become more aware of the movement of people and cameras.

Once the final rehearsal begins, all nonproduction personnel should be asked to leave the studio and observe the production from the monitor in the conference room.

Of course, depending upon the type of show being produced, these final rehearsals may be selected elements of the show or the entire production. In either instance it is not unwise to record the rehearsals in case certain portions are better performed than during the final taping and you may wish to make certain edits to improve the tape.

When checking audio levels, it is not enough to ask for a mere recitation of "one-two-three-four." The performer has a tendency to lean into the microphone, and the words are unnatural. It is better if the audio levels are checked with a few seconds of the actual words the performer will say, if they are prepared remarks . . . or a few seconds of, "My name is Harvey Smith. I am general manager of the XYZ corporation. Today I am going to . . ."

If the narrator or any performer must take an audio cue, it should be rehearsed so the speaker can get a better sense of timing. For example, in the opening as the music fades and the host begins to introduce the panel of speakers, a smoother transition can be achieved if the host begins to speak at a set level of music rather than taking a hand signal from a camera operator. The speaker can mentally adjust when to start speaking rather than have to start from an abrupt signal.

One last look at the performers' appearance — hair straight, no shiny foreheads, tie not crooked — and taping of the show can begin.

A description of how a director operates in the control booth is found in Chapter 8 (TV Direction). If the show runs smoothly, an immediate review will be all that remains. However, during the taping of the show, with the director's attention focused simultaneously on so many items, it is quite possible that minor flaws have crept in. These could include unclear statements, distracting movement, missed shots, or forgotten superimpositions. Electronic problems in the VTR may have occurred which would not be detected until playback.

No matter how satisfied the director is with the immediate taping results, it is advisable to have everyone stand by until the tape is reviewed and approved.

Prior to review, the director might anticipate some minor corrections and tape a few standard shots that can be used as inserts to cover errors or add variety to sequences that have been permitted to run too long. The shots may include:

1. the total group with their attention directed toward a person or object;
2. one-shots of the individuals looking into the camera;
3. one-shots and two-shots of the individuals listening to other speakers;
4. one-shots of individuals listening while their names are being superimposed;
5. medium shots of visuals, models, illustrations, and so forth;
6. close-ups of objects, numbers, and so forth.

The crew can then cap the lenses of their cameras, kill the studio lights, and take a break, while the producer, director, and participants meet in the conference room to view a playback.

A complete review, without stops, is advised so as to get a feeling of the continuity and pace of the show. In this manner, you can determine if the total presentation hangs together, and if major revisions for content or impact are necessary. Assuming that the basic show says what you want it to say in an acceptable time, then specific sequences on the tape can be played back to see if they need refinement.

The director must advise the participants that tape revisions are not easily performed, and that any picture revision, no matter how slight, is an expense item and means a degradation in the quality of the final tape. The participants should also be advised that they will be more aware of errors than the viewing audience, and that certain flaws are sure to go unnoticed.

Should any conflicts arise regarding reshooting or editing, the final decision is up to the producer, who must worry about results and budgets. (These types of conflicts don't normally arise in a small-studio production, but occasionally a performer will insist upon a word change or the deletion of a shot that he or she feels is unflattering.)

After the review, the necessary sequences are reshot . . . and/or everyone is congratulated and thanked for having cooperated.

The equipment is covered; the sets and properties are put away; and (in case we haven't mentioned it before) the studio and control room are cleaned.

CHAPTER 14

SCRIPT PREPARATION

In most small-studio productions you will be working without a script. You will instead be using a combination of diagrams, working notes, and elaborate cues to guide the performers, the cameramen, and yourself. Your productions will be more casual and conversational.

One should, however, gain some experience with scriptwriting, if only to see the total complexity of production. A properly prepared script covers all the bases; it investigates the use of every method or technique available to tell the television story.

Where to Start. You begin the process of writing a script by a consideration of the most important ingredient of the communications process — the audience. It is absolutely essential for you to put consideration of content well aside, at least until you've answered some important questions about the needs of the people whom you expect to watch the production.

You may find it useful to write down everything you know about the audience. Only by a careful examination of all these facts can you begin to assure yourself that your production is going to be successful. You must produce selected messages for selected audiences.

Who *is* the audience? Where are they now? Why should they want to watch your production? Have you ever talked to segments of this particular audience? When must you communicate with them? Is there a deadline for the usefulness of your message? What do they need to know in order to understand your production? Why should they get involved with your message? What benefits are going to accrue to them? How long will your message to them be timely? Did *they* ask you for the production? Did somebody else ask for it?

Once you are satisfied you understand *who* the audience is, and why you're doing the production (and when, and where, etc.), the same process must be undertaken for the material itself.

You must ask questions about the material you intend to produce. What is the problem? Why are we doing it? When did it happen? How frequently does it happen? How much? How many? What happens if we don't tell somebody about it? When can we get it done? Where do we have to go to do it? Who needs to be in it? And so forth.

But those are only theoretical guides. The only way you can gain an appreciation for script planning or production is to get involved in the real process, to produce something that is yours — and to do it by yourself.

We recommend that you work on a production in a way that becomes more than just an exercise. We suggest you consider doing a small-studio video tape production about *your own abilities and capabilities for doing small-studio video tape production.*

We'll do part of the work for you — the easy part. You can do all the hard work of throwing out information you don't need, making judgements, and doing the actual labor involved.

1. Who is the audience? There are several answers to this question. The first three should include the following: A part of the audience is that segment of management that approved the purchase of the equipment (and hired *you*) in the first place. They would be predisposed to "like" anything you did by way of explaining what the equipment is, how it works, and what you intend to do with it. They might even be grateful. *They* would benefit. It would be useful to *them.* It would tend to support their original contention that television can be, or should be, a useful communications tool.

A second group that would benefit would be those members of the organization who didn't know what was going on, but would like to be in on it. They may be a little distrustful and suspicious. A forthright presentation about *what* you're doing and *why* would be psychologically very helpful, both from your standpoint and theirs. It would serve as an invitation for them to become a part of the effort — to conceive of ways of *using* this communications tool — to make them a part of it.

A third group who could be helped enormously includes all the people who, at one time or another, will use the facilities. Some of these people are already very anxious to participate. Others will be apprehensive. A production about what your facilities will do can serve to educate them while it makes them more comfortable.

2. *Why is the production being done?* To satisfy a variety of audience needs, and to meet some of your own needs. The production will serve to reassure, inform, and raise the expectations of your audience. It will serve to tell your story on exactly your own terms. *You* control the content, and *you* control the environment in which it will be seen.

3. *How long should it be?* One of the most time-wasting things you can do is try to estimate the length of a story before you know who the audience is or what the content is. Once you find out these things, the production will tell *you* about how long it should be.

There are dozens of other questions along this line. You should explore as many as possible.

We'll let you make up your own list of subsequent questions. Right now we'll simply present a few more answers.

The *idea* of doing a production about your own (and your studio's) capabilities is absolutely foolproof. Nobody could argue the usefulness of such a production, or the common sense of its being your *first* production. It will, after all, *test* the adequacy of the equipment, the utility of the studio, and be of invaluable assistance to the production crew . . . a sort of proving ground for the necessary teamwork.

It really can serve to put potential performers at ease, particularly since they will view the tape in your own studio and be able to see both the television version and the real version of the same things (space, equipment, etc.).

Such a production will be used for a long time, and shouldn't quickly go out of style. Should you gain more experience in a particular area you can reshoot sequences and re-edit the original version.

A successful production will save you a lot of time. The screening of your production on capabilities could conceivably become a part of VIP tours of the company as a whole. You will find it a good deal easier to thread up a video tape, than to make your own *live* presentation of the goals, objectives, and capabilities of the facility and yourself.

The following is only a model. You can handle the material any way you want, in any order you want. You will have to amend the script anyway, to make it serve your specific needs.

Here are some general tips: wherever possible think in terms of pictures rather than words, or at least begin to think of interesting ways of integrating the two.

Let's begin by considering the *least* bothersome aspect of script preparation — the *form* or *format* itself. Other things about content and organization will occur to you as we go along (not in simply reading this chapter, of course, but when you begin your own script for your own production).

The basic form of a TV or film script is an 8-1/2″ × 11″ piece of paper divided into two vertical halves. The left-hand side of the sheet is reserved for information about pictures. It is usually typed in initial caps and lower case. The right-hand side discusses the *words* used by the narrator or the performers to tell the story. This information is usually prepared in all caps (easier to read).

We will proceed directly to the practical matter of your script — the one that you will use (and by using it better understand the problems of planning and production). The following script will be used in the production of a TV tape to be produced with a basic two-camera system with a switcher-fader (to cut or dissolve, where necessary, between shots).

VIDEO	AUDIO
In black	(Music up)
Camera 1 on title card in studio. Fade in on card: "(Your organization's name) Television Capabilities"	
MS, limbo shot of narrator. Head and shoulders only. (Camera 2)	THE PURPOSE OF THIS PROGRAM IS TO EXPLAIN OUR TELEVISION CAPABILITIES, AND THE BEST WAY TO DO THIS IS TO LOOK AT WHAT REALLY GOES ON . . . BY WHOM . . . WITH WHAT . . . AND WHERE.
	BY DOING THIS WE MAY ELIMINATE MOST OF THE BLACK MAGIC THAT GOES ALONG WITH OUR JOB . . . BUT WE'RE WILLING TO TAKE THE RISK BECAUSE TELEVISION IS THE MEDIUM OF TODAY.

VIDEO	AUDIO
	WE'RE TALKING ABOUT A WAY THAT CAN MEET THE NEEDS OF COMMUNICATION HEAD-ON . . . A TECHNIQUE THAT CAN REFINE, CLARIFY, AND SIMPLIFY SOME OF THE INFORMATION WE ARE CURRENTLY TRANSMITTING.
Dissolve to MCU, Camera 1. This shot is on the tight end of the zoom. Slowly zoom out, keeping the narrator in the center of the picture.	WE EXPECT TO SHOW YOU HOW TELEVISION WORKS, AND HOW WE INTEND TO USE IT. BUT THE MAIN THING WE WILL DO, IS FORCE YOU . . . THE POTENTIAL COMMUNICATOR . . . TO DO A VERY IMPORTANT THING . . . SOMETHING YOU MAY NOT HAVE DONE FOR A LONG TIME.
As the camera is zooming out, the studio lights are very slowly coming on . . . pulling the narrator out of limbo and placing him or her in an empty studio.	WE WILL FORCE YOU TO CONSIDER YOUR AUDIENCE A LITTLE MORE CAREFULLY. TO THINK ABOUT THE INDIVIDUAL MEMBERS IN THAT AUDIENCE . . . AND TO WONDER ABOUT WAYS TO COMMUNICATE WITH THEM IN SIMPLER, FASTER, AND MORE DIRECT WAYS.
	YOU CAN'T STAND IN FRONT OF A CAMERA AND READ A REPORT. YOU HAVE TO LOOK AT YOUR AUDIENCE SQUARE IN THE EYE AND TELL A STORY. ONE THAT MAKES SENSE. AND ONE THAT THEY CAN UNDERSTAND. AND ONE THAT THEY CAN *SEE*.
Still on Camera 1, but now on LS, with studio evenly illuminated.	THIS IS WHERE WE WILL HELP YOU WITH YOUR STORY-TELLING PROBLEM. THIS IS THE STUDIO.

VIDEO	AUDIO
Still on Camera 1, pan as narrator walks to VTR.	AND OVER HERE IS THE REASON FOR THE STUDIO . . . THE VIDEO TAPE RECORDER, CALLED "VTR" FOR SHORT.
Cut to Camera 2 and CU of VTR.	THIS PIECE OF EQUIPMENT WORKS IN THIS STUDIO, BUT CAN ALSO BE MADE TO WORK IN THE PRESIDENT'S OFFICE . . . ON A PRODUCTION LINE . . . IN A LABORATORY . . . EVEN OUT OF DOORS. IT RECORDS BOTH PICTURE AND SOUND.
Cut to Camera 1. MS of VTR and "prop" camera.	THE VTR WORKS IN CONJUNCTION WITH A TELEVISION CAMERA. THE VTR RECEIVES A SIGNAL FROM THE CAMERA WHICH IT TRANSLATES INTO MAGNETIC IMPULSES.
Cut to Camera 2. CU of video tape.	THESE IMPULSES ARE THEN TRANSFERRED ONTO A SENSITIZED TAPE . . . NOT UNLIKE THE TAPE YOU'VE USED FOR YOUR AUDIO TAPE RECORDER.
Cut to Camera 1. MS to include hand that stops machine . . . rewinds tape . . . pushes "Play . . ."	WHEN THE TAPE IS REPLAYED ON THE SAME MACHINE . . . OR ONE BY THE SAME MANUFACTURER . . . IT BECOMES A VIDEO TAPE PROGRAM . . . THAT CAN BE VIEWED ON A STANDARD TELEVISION RECEIVER.
Zoom out to LS to include screen.	
Dissolve to LS of studio. Camera 2.	BUT BEFORE WE TALK ABOUT THE FLEXIBILITY OF VIDEO TAPE RECORDING, LET'S TAKE A LOOK AT THE STUDIO AND WHAT IT CONTAINS . . . AND THE

VIDEO	AUDIO
	VARIOUS THINGS THAT CAN BE DONE IN IT. THE BASIC STUDIO ISN'T VERY LARGE. THE DIMENSIONS ARE ____BY____. THE CEILING HAS A HEIGHT OF____FEET TO ACCOMMODATE THE LIGHTS THAT MUST BE USED TO PROVIDE AN ADEQUATE LEVEL OF ILLUMINATION.
Camera follows narrator to center of studio set.	LET'S TALK ABOUT LIGHTING FOR A MOMENT SO THAT YOU WILL UNDERSTAND (BY ACTUALLY SEEING) WHY CERTAIN LEVELS AND TYPES OF LIGHT ARE NEEDED.
Cut to Camera 1. MLS, dim view of narrator. Light levels down.	YOU NOW SEE ME STANDING IN THE STUDIO WITH LIGHT AT THE LEVEL THAT IS NORMALLY FOUND IN AN OFFICE OR A CLASSROOM. THE PICTURE LOOKS THE WAY IT DOES BECAUSE THE *SYSTEM* IS STRAINING TO SEE ME. AS WE TURN ON MORE LIGHTS, THE PICTURE IMPROVES.
Zoom in to MCU, head-and-shoulders of the narrator.	THE MORE LIGHT (UP TO A CERTAIN POINT) THE BETTER THE PICTURE. SO YOU JUST HAVE TO LEARN TO ACCEPT THE HIGH LIGHT LEVELS YOU WILL FIND WHEN YOU ARE A PARTICIPANT IN A PROGRAM. IT MAY BE SLIGHTLY UNCOMFORTABLE FOR A WHILE, BUT THE INCONVENIENCE IS MORE THAN OFFSET BY YOUR IMPROVED

VIDEO	AUDIO
	APPEARANCE IN THE FINAL PRODUCTION.
Cut to Camera 2. LS of person silhouetted by light behind him.	FIRST, LOOK AT THE BASIC LIGHTS WE USE TO LIGHT *THE SET*. HERE THEY ARE. IT IS IMPORTANT TO PLACE LIGHT ON THE BACK WALLS SO THEY WON'T LOOK DINGY AND SHADOWY.
MS talent with hair light.	NEXT IS SPECIALIZED ILLUMINATION CALLED A *HAIR LIGHT*. IT IS AN ACCURATE DESCRIPTION OF WHAT IT DOES . . . IT FALLS DOWNWARD ON THE PERFORMER AND SERVES TO HIGHLIGHT THE TOP OF THE HEAD (OR THE HAIR) AND THE TOPS OF THE SHOULDERS SO THAT THE PERFORMER SEEMS TO HAVE DIMENSION AND IS SEPARATED FROM THE BACKGROUND.
Dissolve to Camera 1. MLS of talent with key light.	THE NEXT IMPORTANT LIGHT IS THE *KEY LIGHT*. THIS PROVIDES BASIC ILLUMINATION FOR THE PERFORMER. WITH IT, AND THE HAIR LIGHT, THERE ARE CURVES AND ANGLES APPARENT IN THE FACE AND BODY.
Dissolve to Camera 2. MS, different view, adding studio fill light.	FINALLY, WE HAVE *FILL LIGHT*, THE GENERAL, OVERALL ILLUMINATION FOR THE SET, AND THE PEOPLE AND THINGS *ON* THE SET.
MCU, head and shoulders. All lights on. Cut to Camera 1. MCU, of "prop" camera. CU, talley light.	TAKEN ALL TOGETHER, THE PICTURE IS A PLEASING ONE. WE'LL NOW TALK ABOUT CAMERAS. HERE IS ONE USED IN

VIDEO	AUDIO
	THIS FACILITY. IT IS A LIGHT-WEIGHT (brand name) CAMERA EQUIPPED WITH A TALLEY LIGHT . . . SO THE PERFORMER CAN TELL WHICH CAMERA IS ON . . .
"Prop" camera turns MCU, viewfinder.	WITH A VIEWFINDER . . . SO THE CAMERAMAN CAN SEE WHAT THE LENS IS SEEING . . .
"Prop" camera turns MCU, lens or lenses.	AND A ZOOM LENS (OR A COM-PLEMENT OF____ LENSES).
Cut to Camera 2. LS of studio.	THIS IS THE EFFECT OF A ZOOM LENS (OR THE INDIVIDUAL LENSES) WHEN USED IN VARIOUS PARTS OF THE STUDIO.
	FROM THE EXTREME END OF THE STUDIO A ZOOM LENS CAN SEE THIS MUCH OF THE SET, BUT CAN ZOOM INTO THIS.
From LS of studio, zoom into ECU of portion of narrator's face.	IF THE CAMERA IS QUITE CLOSE TO THE PERFORMER, IT CAN ZOOM INTO A SHOT THAT LOOKS LIKE *THIS*.
Cut to Camera 1. LS, studio and narrator.	FOR PURPOSES OF IDENTIFICA-TION, IN THIS STUDIO, THE FOLLOWING CLASSIFICATIONS ARE GIVEN TO DIFFERENT KINDS OF SHOTS.
LS, same.	THIS SHOT, THE BROADEST COVERAGE POSSIBLE IN THIS STUDIO, IS CALLED A *LONG SHOT*. IT INCLUDES ME AND A CONSIDERABLE PORTION OF THE SET.
CU, narrator.	THIS SHOT, CONSIDERABLY CLOSER, IS CALLED A CLOSE-UP.
MS, narrator and set.	A SHOT MIDWAY BETWEEN THE

VIDEO	AUDIO
	TWO (BETWEEN A LONG SHOT AND A CLOSE-UP) IS CALLED A MEDIUM SHOT.
Zoom into ECU of eye.	FOR THE ULTIMATE IN DETAIL IT MAY BE NECESSARY OR USEFUL TO GO TO AN *EXTREME* CLOSE-UP.
Cut to Camera 2. MS of narrator.	SO THOSE ARE THE SHOTS. IT IS ALSO OF GREAT IMPORTANCE TO KNOW HOW TO GET FROM ONE SHOT TO ANOTHER.
Cut to Camera 1 on other talent. LS.	ONE WAY, THE EASIEST, TO GET FROM ONE SHOT TO ANOTHER IS TO *CUT* . . . LIKE THIS.
Cut to Camera 2.	AND THEN BACK AGAIN. A COMFORTABLE WAY TO GET FROM ONE SCENE TO THE NEXT IS THIS WAY.
Fade in to Camera 1 on narrator. Fade out to Camera 2. Fade in to Camera 1. MCU of narrator.	THIS IS AN EXAMPLE OF A FADE. FADE IN. FADE OUT. FADE IN.
Fade from Camera 1 to black, and into a different shot . . . with Camera 2. CU.	TO SHOW TRANSITION OR PASSAGE OF TIME IT IS POSSIBLE TO FADE TO BLACK AND FADE BACK INTO THE NEXT SCENE . . . AS IN THIS ILLUSTRATION.
Dissolve to Camera 1 with narrator over LS of other talent.	ANOTHER VERY COMMON TECHNIQUE IS TO SUPERIMPOSE THE SECOND PICTURE ON TOP OF THE FIRST PICTURE. HERE IS AN EXAMPLE OF A FAST DISSOLVE.
Dissolve from Camera 1 to Camera 2.	HERE IS A SLOW DISSOLVE.
Cut to Camera 1. MCU of narrator.	THIS IS ANOTHER PRACTICAL REASON FOR HAVING A DISSOLVE THAT IS SPLIT BETWEEN ONE PICTURE AND ANOTHER.

VIDEO	AUDIO
Super name with Camera 2.	I'M SURE YOU'RE ALL FAMILIAR WITH THIS EFFECT. THE NAME OR TITLE OF A PERSON APPEARING WHILE THAT PERSON IS ON THE SCREEN.
Fade out Camera 1, leaving only super. Zoom out to show basic set-up.	THIS IS ACCOMPLISHED IN THE FOLLOWING WAY: . . . WE'LL ZOOM OUT WITH THE CAMERA THAT IS SHOOTING THE SUPER TITLE SO THAT THE STAND, AND THE END OF THE STUDIO, ARE VISIBLE.
Cut to Camera 1. CU of narrator. Fade in super.	NOW, HERE IS THE PICTURE UNDER WHICH THE TITLE WILL APPEAR. SLOWLY THE TITLE IS SUPERED IN, UNTIL THE PROPER EFFECT IS ACHIEVED.
Zoom out to MS of narrator.	SO THOSE ARE THE BASIC SHOTS AND THE WAYS OF PUTTING THE SHOTS TOGETHER.
Hold shot. Return lighting to original limbo shot.	ALL RIGHT, THAT'S THE EQUIPMENT . . . HOW IT WORKS . . . WHERE IT'S OPERATED . . . AND SOME OF THE BASIC TECHNIQUES. BUT THAT'S ONLY *HALF* OF THE OPERATION. THE OTHER HALF IS WHAT IS BEING SAID AND WHY IT IS BEING SAID.
Cut to Camera 2. CU at different angle.	THE PRIMARY REASON FOR USING THE TELEVISION MEDIUM IS THAT WE WISH TO VISUALLY COMMUNICATE SOME INFORMATION. WE ARE ALL AWARE OF THE IMPACT THAT IS INHERENT IN A TELEVISED MESSAGE.

VIDEO	AUDIO
	WE USE TELEVISION TO SELL (OUR) PRODUCTS . . .
Dissolve to film — use television commercial of company (or any standard commercial).	
Dissolve to Camera 1. MS of narrator.	WE USE TELEVISION TO SHOW US PLACES OR SCENES WE WOULDN'T ORDINARILY SEE . . .
Dissolve to film of another facility (or a sales situation, or a lab test).	
Dissolve to Camera 1. MS of narrator.	WE ALSO USE TELEVISION TO BRING US PEOPLE WE WOULDN'T ORDINARILY SEE OR HEAR . . .
Dissolve to film . . . Show company president (or show national figure, or outside consultant).	
Dissolve to Camera 1. MS of narrator.	WE AT (name of organization) WILL USE TELEVISION TO EXPAND OUR COMMUNICATIONS, MARKETING, AND TRAINING PROGRAMS. HERE IS SOME SAMPLE PROGRAMMING.
Fade out narrator. Fade in taped excerpts from programming. (Run for about three minutes.)	
Fade in Camera 1. MS of narrator in limbo setting. Cut to Camera 2. CU of narrator at different angle.	AS YOU SAW IN THESE EXCERPTS FROM SOME OF OUR PROGRAMMING, EACH WAS A *VISUAL* PRESENTATION OF A SUBJECT THAT WAS OF CONCERN TO SPECIFIC GROUPS AMONG OUR CUSTOMERS AND OUR EMPLOYEES.

VIDEO	AUDIO
	BUT IN EACH CASE THE PROGRAMMING CREATED AND MAINTAINED INTEREST BECAUSE IT *INVOLVED* THE AUDIENCE.
Start slow zoom out.	EACH SEQUENCE WAS DESIGNED AND PRODUCED TO MAKE ITS OWN POINT. THE *SERIES* OF SEQUENCES WAS ALSO CONSTRUCTED TO MAKE AN EVEN BROADER POINT.
Continue and end on LS of narrator.	THE RESULTING SERIES IS A PROGRAM THAT IS PART OF A TOTAL COMMUNICATION—IT CREATES INTEREST, IT TELLS THE AUDIENCE ITS PURPOSE, IT GIVES THEM THE INFORMATION, AND IT TELLS THEM WHAT IS EXPECTED OF THEM.
Cut to Camera 1. MS of still photos that describe other media. Pan L to R. Zoom in to CU of each photo. Pan R to L.	THE VIDEO COMMUNICATION MAY BE PART OF A STILL LARGER COMMUNICATION THAT INVOLVES BROCHURES – HANDBOOKS – TRAINING MANUALS – CATALOGS – OR ON-THE-JOB TRAINING WITH ACTUAL EQUIPMENT OR MODELS.
Zoom out to MS of still-photo board.	BUT IT'S ALL FOR THE PURPOSE OF GETTING INFORMATION TO A SPECIFIED AUDIENCE – FOR A DESIRED RESPONSE.
Dissolve to Camera 2. MS of narrator. Do slow zoom in. End on CU during his last line of narration.	AND THAT'S WHAT WE'RE ASKING FROM YOU. THINK IN TERMS OF TELEVISION. THINK IN TERMS OF A VISUAL MESSAGE. THINK IN TERMS OF A TOTAL COMMUNICATIONS PACKAGE.

VIDEO	AUDIO
	YOU DEVELOP THE MESSAGE – *WE'LL* SEE THAT IT GETS DELIVERED.
Repeat silent version of sequence of excerpts with music under.	
Fade in credits. Fade in logo. Fade to black.	(Music under)

CHAPTER 15

LOCATION SHOOTING

As we indicated in Chapter 6 (Production Planning, describing the use of a TV production planning sheet) we consider location production to be of two kinds, in-house or off-site.

In-house means that shooting will be done in a location other than the studio facility. This might mean a laboratory, a classroom, an assembly line, a test facility, an auditorium, a meeting, a conference, etc. It supposes, however, that the physical problem of moving equipment and material won't be a strenuous one.

Off-site production implies the physical movement of equipment, materials, supplies, and people, to a fairly distant point. The problems of moving delicate equipment across a major city are just as complicated as those of moving the same equipment over a much longer distance. The equipment needs to be packed properly and cared for.

Despite the claims of some manufacturers, television cameras and recorders do not stand up well during travel and extensive packing and unpacking. If it is necessary to shoot a show at a location other than where your facility is situated, it is advisable to check first to see if rental equipment is available in the area. The two main sources for equipment are nonbroadcast production houses, which can also supply personnel . . . and franchised audio/visual dealers for major VTR manufacturers. It is unlikely that other nonbroadcast television producers located in the area of the shooting will wish to rent their equipment for an off-premise production, but they may be worth a few inquiries if the first two sources are unavailable.

Nonbroadcast production companies are discussed more fully in the next chapter (Using Outside Facilities). The few such companies that are in operation are in major cities, but most are willing to send out crews and equipment to other locations. Their rates may appear, at first, to be a high price for personnel and equipment that you are

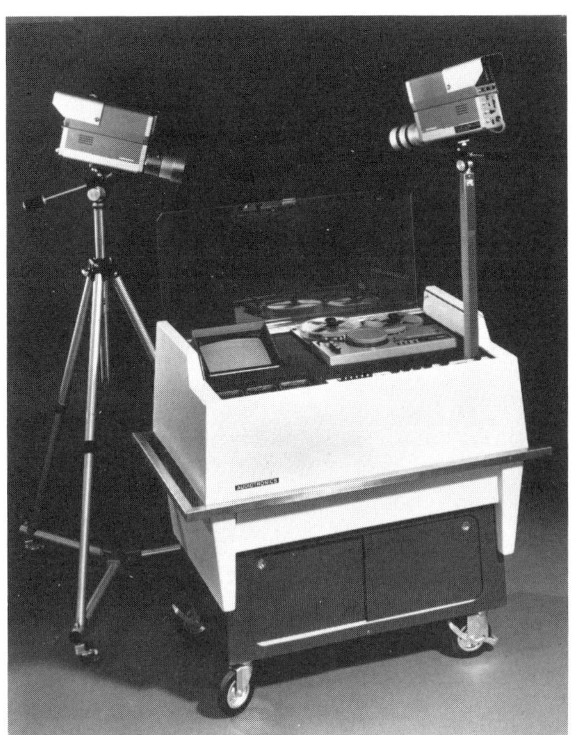

(Photograph courtesy Audiotronics Corporation)

Complete TV studio on wheels — specially adapted for location shooting

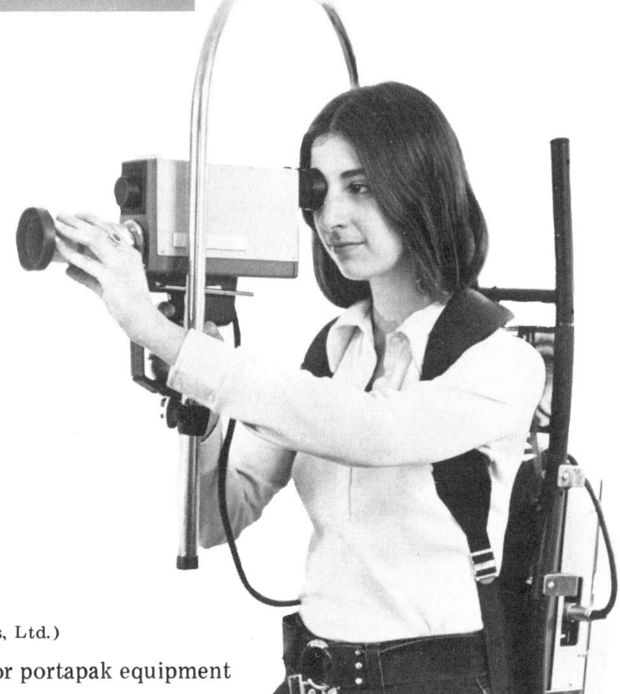

(Photograph courtesy K and H Products, Ltd.)

Backpack brace and support arm for portapak equipment

also able to supply. But the final cost may end up as savings from a more efficient operation and undamaged equipment. If a production company using 1-inch tape is available, a two-camera crew and VTR operator will cost between $900 and $1500 a day plus tape and travel expenses.

There is still another alternative, and that is to use a remote crew from a local television studio or commercial broadcast company which would tape the show on 2-inch tape from which a 1-inch master could be dubbed. Remote units using a 2-inch format have a minimum rental cost of $3000 a day, and in most cases the costs run higher because of travel time, tape costs, and postproduction on quad equipment. If there are no budget restraints, this is one way of producing your show on location, but, in this case, it *does* cost more to go first-class.

The reasons for location shooting are varied and important:

1. Training tapes where the equipment being demonstrated is permanently installed or too large for the studio.
2. Highlights of sales meetings or seminars of interest to other personnel.
3. Meetings of stockholders, management, or regional representatives that can be shown to other company personnel.
4. Results of laboratory or field tests.
5. Athletic events or outdoor demonstrations.
6. Openings of new facilities or other events of note.
7. For educational purposes, useful interviews, and demonstrations.

Whether the location of the shooting is in-house or off-site, it will go more smoothly if it is planned well in advance.

The first step in location shooting, of course, is to inspect the site (even if you've shot there on previous occasions). The physical aspects of the room will determine the amount of equipment you can utilize and the types and lengths of lenses that will be needed.

For example, if you wish to record a sales or stockholder meeting being held in the grand ballroom of a hotel, the balcony of the room would appear to be an ideal spot to position the television equipment. However, a quick check of distance from balcony to stage will indicate that special long lenses will be needed and even these would still not be able to provide close-up shots of the speakers. It would be better to find a closer spot to the stage, even if it means having to contend with some traffic.

In the same manner, shooting in small offices or laboratories requires wide-angle lenses. In addition, the available space may be too cramped for the VTR and technical operators to work in the immediate area. Extra video and audio cables will have to be provided.

Probably the major difficulty in location shooting is obtaining proper light levels. As we have indicated in previous chapters, the degree of contrast and the depth of field of the video picture are dependent upon the amount of light that can be placed on the subject. In most instances with location shooting you will be restricted to shooting with just the available light, or, at best, be able to bring in a few stand lights.

Despite the handicaps that insufficient light present to you and the quality of the recording, it is probably not advisable to bring in additional lights to a large-scale meeting you are recording. You should keep in mind that the meeting is basically being held for the benefit of the immediate audience, and that television lights can only distract from what is being presented on stage. You can get additional illumination by bringing more of the stage lights to bear on the participants, or by getting a follow spotlight to focus in on the speaker's stand or panel table.

There is usually no problem in bringing portable lights to an office, factory, laboratory or classroom. Your main concern in those instances will center around adequate space and the heat that will be generated.

The placement of the VTR equipment is also an item which must be determined. The director, who may wish to communicate over the headsets with the cameramen during the shooting, will have to be separated from them by a wall or a room. In some cases this will not be possible, and a series of hand signals will have to be worked out so the director can indicate the type of shots desired and when the shot is about to be taken.

There will be many instances in location shooting when the director will also operate the camera, such as in the case of interviewing a top company salesman in his office or while he is calling on customers. This situation and similar ones call for a minimum of equipment—maybe some of it hand-held. But the results need not be inferior. Some of the most interesting candid shots are those that are achieved with lightweight, portable equipment.

Location shooting may sometimes offer the possibility of using a following (or follow) shot. If the equipment is portable, this is no

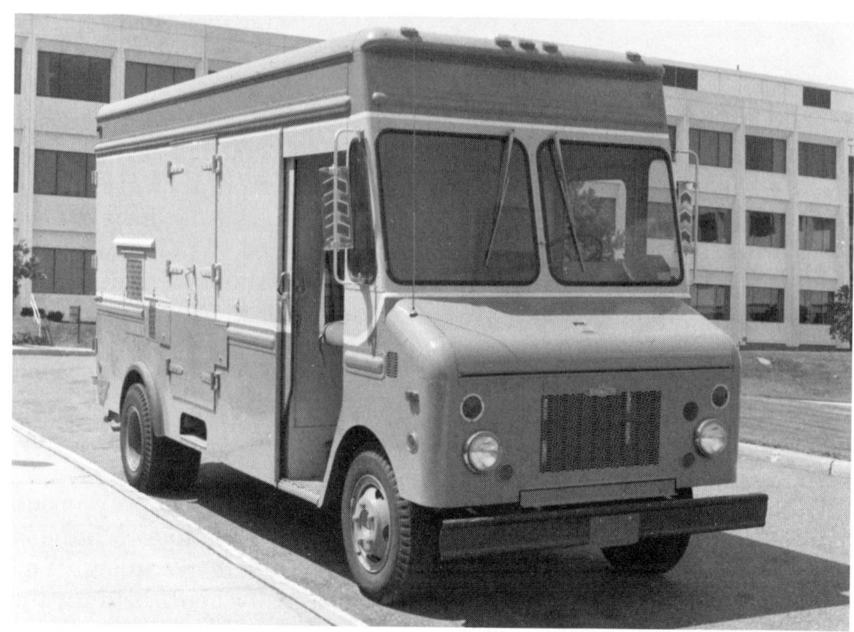

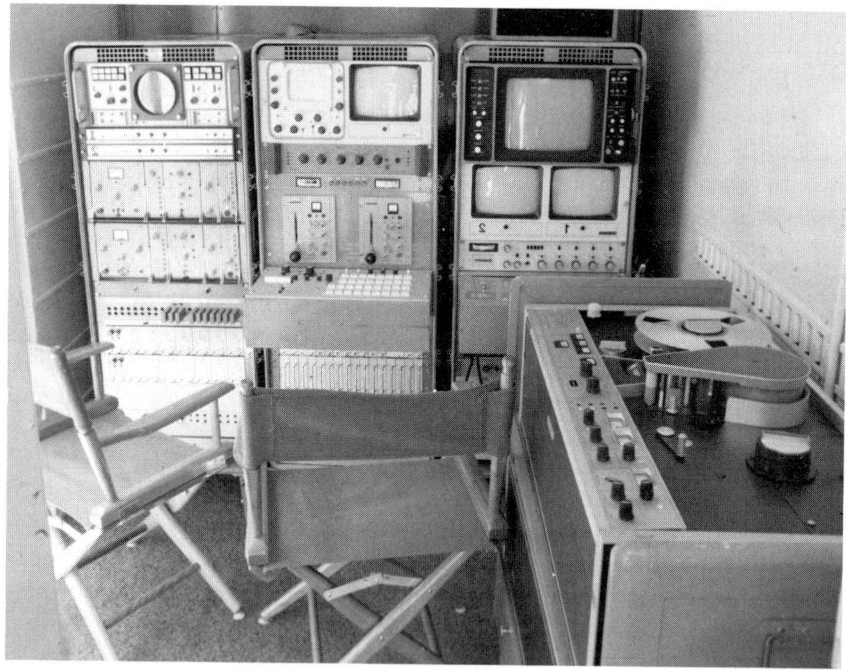

(Photograph courtesy International Business Machines Corp.)

Exterior and interior of mobile unit equipped with 1-inch video tape equipment

problem, and it can even be done where there is enough camera or power cable to enable the camera to move a reasonable distance.

The function of a following shot is to move along with the moving performer or the subject matter. The distance between the camera and the subject remains about the same, and the camera may be placed in front of, alongside, or behind the subject. The object is to keep the shot focused on the subject, with no regard for the background that moves along behind the shot.

A shot such as this offers a great deal of variety and can be used to good effect particularly in industrial situations where there is considerable activity. An example of the use of such a shot would be to follow an item as it moves along a conveyor belt. A greater sense of movement is shown by a following shot — much more so than a static shot that watches material moving past the camera.

The main problem with a following (sometimes called trucking) shot is finding the wheeled vehicle to place the camera on. A variety of things can be tried, including electric carts and trucks often found in manufacturing areas.

When shooting off-site, you may encounter union problems not experienced in your own facility. For example, most large meetings or exhibitions are held in hotels or convention halls that have union contracts. This means that a union electrician must hook you up to a power source; union men must handle both the lights and the sound. And even though there is currently no union that specifically handles nonbroadcast television equipment and recording, you may have to hire a stand-by person (or two) while you perform the work. From our experience, we have gone ahead and made television set-ups on the assumption that the union personnel at the site will let their presence be known, and will negotiate their requirements at that time. This is usually what has happened — and we have encountered no major difficulties in working out a reasonable solution.

We would like to introduce another worksheet entitled "TV Location Production." As with the other planning sheet, there will be items you may wish to delete as unnecessary to the operation you have, or propose to have. You may also wish to add elements that we have not added, specialized requirements that are a function of the type of work you do the most of, or needs of the subject itself. In some production cases, there will be need for specialized lighting instruments, lenses, and various kinds of support equipment. If these are used, then they should be indicated on a sheet designed to expedite location production.

A sheet should be used, if only to keep track of the equipment that will be needed on the job. Nothing is more frustrating than to go to the considerable expense (in time and money) of setting up, only to find that you are missing some seemingly inconsequential, but altogether necessary component.

This sheet can be used for three purposes: to aid in estimating requirements of time and money; to assist in planning location shooting; and as a checklist of materials that you must remember to take along in addition to what you are renting. (Never assume that certain materials, equipment, or resources are going to be at the other end, unless you have absolute confirmation that they are available. It is best to bring your own slides, hand props, extra lenses.)

1. **The date, production number, and title** of program are all self-explanatory.

2. **Location(s).** It is important to indicate precisely the location (plant site, room number, lab location, classroom, etc.) where the shooting is to take place. If more than one location is involved, then additional sheets should be used, since requirements may differ from one room or location to another.

3. **Participants.** The name of the performer(s) should be noted, if only to let him or her know that this production involves other persons and a great deal of preparation.

4. **Date of arrival.** It is very important that everybody connected with the production (and the location) be informed about your intended work. Otherwise a lot of confusion and considerable loss of time will result. It is sometimes advisable to arrive before the actual day of shooting to set up backgrounds, move furniture, and confirm that long-distance communications have been received and acted upon.

5. **Estimated time required.** This is equally as important as the date of arrival. People at the other end need to know your requirements in order to prepare themselves for the down-time that will be incurred, and to meet the special needs of the situation. If, through poor planning or organization, production requires considerably more time than was anticipated, a lot of money could be at stake. Once again, it is important that everybody concerned "get the word."

6. **Additional personnel required.** It is very often the case in location situations that additional manpower is useful for purposes of unloading, setting up, and so forth. It is often very useful, and sometimes required by unions, to have an electrician working with you on location.

TV LOCATION PRODUCTION	1.) DATE:	PRODUCTION #
	TITLE:	

2.) LOCATION(S):

3.) PARTICIPANTS:

4.) DATE OF ARRIVAL: DATE OF SHOOTING:

5.) ESTIMATED TIME REQUIRED:

6.) ADDITIONAL PERSONNEL REQUIRED:

7.) ELECTRICAL REQUIREMENTS:

8.) OTHER EQUIPMENT OR MATERIALS REQUIRED:

9.) EQUIPMENT TO BE TRANSPORTED (T) OR RENTED (R)

RECORDER	____	MICROPHONES	____
CAMERAS	____	MIKE CABLES	____
TRIPODS	____	FADER/SWITCHER	____
HEADSETS	____	AUDIO RECORDER	____
CLEANING MATERIALS	____	POWER CABLES	____
LIGHTS	____	JUNCTION BOXES	____
LIGHT STANDS	____	SPARE BULBS	____

10.) MATERIALS TO BE TRANSPORTED:

VIDEO TAPE	____	PROPERTIES	____
VISUALS	____	AUDIO TAPE	____
		MAKE-UP	____

11.) TYPE OF TRANSPORTATION:

12.) INFORMATION ON ACCOMMODATIONS:

13.) PERSON IN CHARGE: (TV PRODUCTION) ____________ (ON-SITE) ____________

MISC. INFORMATION

14.) COPIES FORWARDED TO:

7. Electrical requirements. As indicated, this area can pose some problems away from home. These problems are best solved by somebody familiar with the electrical distribution systems of the facilities you're working in. Since you are certain of using considerable amounts of power for your equipment, cameras, and lights, you would do well to communicate your electrical requirements well ahead of time so that on-location electricians can accommodate your needs comfortably (instead of being faced with a last-minute crisis situation such as blowing out half of the fuses in the place).

8. Other equipment or materials required. This entry refers only to the equipment or materials you expect to be provided by the facility where you're shooting. *Your* additional equipment is covered under rental equipment needed. The on-site requirements could include ladders, furniture, additional lighting equipment, models, samples of product, etc. You should also be alert to the needs of the production crew on location. Special arrangements may have to be made for transportation to and from sites, for meals, etc.

9. Equipment to be transported or rented. This, to be sure, is only a partial list. Be sure to include *everything* that may be needed at the site and indicate whether you are bringing it (T) or renting it (R). As mentioned earlier, what you forget to bring may very well be impossible to acquire on location. Take everything you may need. This includes any necessary tools, replacement parts, repair kits, etc. And be certain to get a confirming letter and equipment list from suppliers of rental equipment.

10. Materials to be transported. This list must be as exhaustive as the other list, and should include an extra copy of the script or outline.

11. Type of transportation. Everybody in the production crew should know how people and equipment are to travel.

12. Information on accommodations. It is important to have detailed information about accommodations, per-diem allowances, and other data of interest to the crew.

13. Person in charge. The entry on responsibility needs no explanation. Obviously, one must be familiar with the people in charge. An abundance of phone numbers should be collected so that people in authority can be reached both at work and at home.

14. Copies forwarded. The last item should also be self-explanatory. Does everyone connected with the production know what's going on?

CHAPTER 16

USING OUTSIDE FACILITIES

In Chapter 1 we stated that the frequency of production, and the ultimate cost per show should be one of the criteria for determining whether to build an in-house studio or whether to use outside facilities for production. Another reason for investigating outside facilities is that a particular show may require a high degree of advanced video techniques which your own equipment may not be capable of handling. For example, a show may require the integration of film, slides, superimpositions, or split-screen images. At that point, it may be time to look for some outside assistance.

Outside television production facilities can be either companies specializing in television tape productions or local television stations that rent their studios and personnel when they are not telecasting or taping their own shows.

Most of the television production companies were designed to tape or film television commercials, or to carry the overload of television network facilities. They are located mainly in major cities, and their equipment rivals that of the major television networks, and will probably far exceed your requirements. Thus, you may have to pay for overproduction capability, but there are some ways that costs can be held down to fit within your limited budget.

There are some companies in the country designed exclusively for nonbroadcast television production. These are companies that deal mainly in 1-inch and 1/2-inch formats. Facility space is kept to a minimum, and the technical personnel are nonunion. Many of these companies were forced out of business as equipment costs came within the reach of most potential users, and when the larger studios began to accommodate the industrial television market during slow periods of commercial production.

Additional sources for nonbroadcast television productions are major corporations with extensive in-house facilities. Many such companies have full-time writer/directors and technical personnel, and, in order to hold overhead costs down, have actively sought either to share their facilities or rent them (and their personnel) to other users.

In the same manner, you will find that local television stations would prefer to rent their studios to nonbroadcast users rather than have their crews idle while on full salary schedules.

There is no question that a producer should go "shopping" when considering the use of an outside facility. There may be as much as a two- or three-times variance in costs for equal production competence. And the lowest price could depend solely upon timing — which facility has idle equipment and an idle crew.

If you are located in a major city, you should probably first check on the availability of a nonbroadcast facility such as a major industrial firm that has extensive equipment and full-time production personnel, or possibly an educational institution or school system utilizing closed-circuit television. Their personnel would be most familiar with your requirements and would be sympathetic to keeping costs to a minimum. Their production standards and technical capabilities can be ascertained by merely reviewing some of their past productions.

The major West Coast and East Coast cities are where one would check on the availability of a television production house. A number of them are equipped with 1-inch equipment and have had some experience with nonbroadcast requirements. Also, the production house can usually scale costs to fit a particular budget by offering a smaller studio, smaller crews, and a more flexible time schedule. On the other hand, a television station must always work around its scheduled programs and its small studios are usually set up for daily telecasts such as news or interview programs; and, therefore, the furniture and sets cannot be rearranged very easily.

Both the production house and the television station are often unionized, so certain minimum crews are automatically required. It has been our experience, however, that there is little "featherbedding" in commercial television production. When you are attempting to do a show with one rehearsal and one run-through, you are only too glad to see the individual assignments carried out with a minimum of confusion and a maximum of understanding. One sure way of keeping costs down for a production is to "wing it" — to block some general camera movements and let the cameramen and director work out the details as the show is being taped. But this *can only be done by professionals* who have had years of experience in doing similar shows.

For the most part, let the production house determine how many people are required to put the show together. The salaries of a television crew usually account for less than half the production costs. In reality, you are paying people merely to get the best and most efficient results from equipment which costs hundreds of thousands of dollars. And production companies, like any other business, do not want to price themselves out of the market and have expensive equipment sitting idly by not being used.

Outside production facilities, with a full crew, may vary in costs from $200 an hour (nonunion) to $300-$500 an hour (union), with a two-hour minimum and pro-rated by the half-hour. In addition, there are costs for time to arrange the setting and lights, and costs after the production for the time it takes to put everything away. Some facilities also charge extra for use of drapes, furniture, and use of conference and viewing rooms.

If the purpose in utilizing an outside facility is for greater technical capability, then be prepared to pay for this. The use of a film or slide chain requires an extra person. So does a complex integration of music and sound effects. And a detailed demonstration of a new engineering product may require an extra camera and someone to run it.

A standard union crew consists of:

Director
Assistant director
Technical director
Camera operators (2 or 3)
Audio director
Video technican
Floor manager
Stage crew (2 or 3)
Lighting director

It is sometimes possible to dispense with the services of a lighting director, especially if you are using a standard setting not requiring any shift of scenery or lights. Such a situation would also mean the need for only one stage-crew member.

In some instances, especially at a production house, the director may not be included as part of the crew. You must then bring in your own union director or negotiate a fee with one of the house's staff personnel.

Extra workers such as make-up people, on-camera talent, and off-camera narrator or announcer are not included in the basic cost and their fees must also be negotiated. The studio usually has people it has dealt with in the past, and can indicate what the services will cost.

One of the problems in using a production house, a problem that is not encountered with a television station, is that the members of the studio crew (director, cameramen, floor manager, etc.) are usually not full-time staff and are called in for each production. Thus, the crew may be determined on the basis of who is available, and despite the fact they are all professionals, they may have never before worked together as a team. This causes delays in getting signals across and trying to anticipate what a director is calling for.

At a broadcast facility, a crew that continually does a news broadcast or interview show can invariably turn out a better program at less cost. A good source for this type of crew is a local educational television station, which has had much experience with talk shows.

It must be kept in mind that an error in production means either that you must start over, or you must stop and correct the error and pick up the show from that point. The latter correction then requires some postproduction editing, which costs about $150 an hour, for a minimum of one hour.

If you are not careful about such incidental costs, or if you do not have your show well planned, your final production costs can go far beyond what you anticipated. For example, special settings may be a nice touch, but they must be constructed at an outside union shop, trucked to the station by union movers, and erected and lighted well in advance of the rehearsal. All this, of course, can almost equal the cost of the actual taping of the show. It is well to look over the sets already available at the facility to see if they can be adapted to your program.

Do not be misled, however, into thinking that what is already available is therefore free for your use. Most production houses charge a rental fee for such things as drapes, desks, and chairs, even though they may already be in the studio. The rationale for the rental fee is that the items were purchased by the production house just for programming — the same as the equipment — and they must be rented on the same basis. It is cautioned, therefore, to have no misunderstanding regarding what is included in the basic studio price, and then negotiate a fixed amount for any of the so-called "extras."

Another cost item at an outside facility is the cost or rental of a 2-inch video tape, if you are not recording directly onto a 1-inch VTR. You are permitted to bring your own video tape for the recording, but you may not wish to purchase a $300 2-inch master tape, especially if it is not going to be erased and rerecorded a

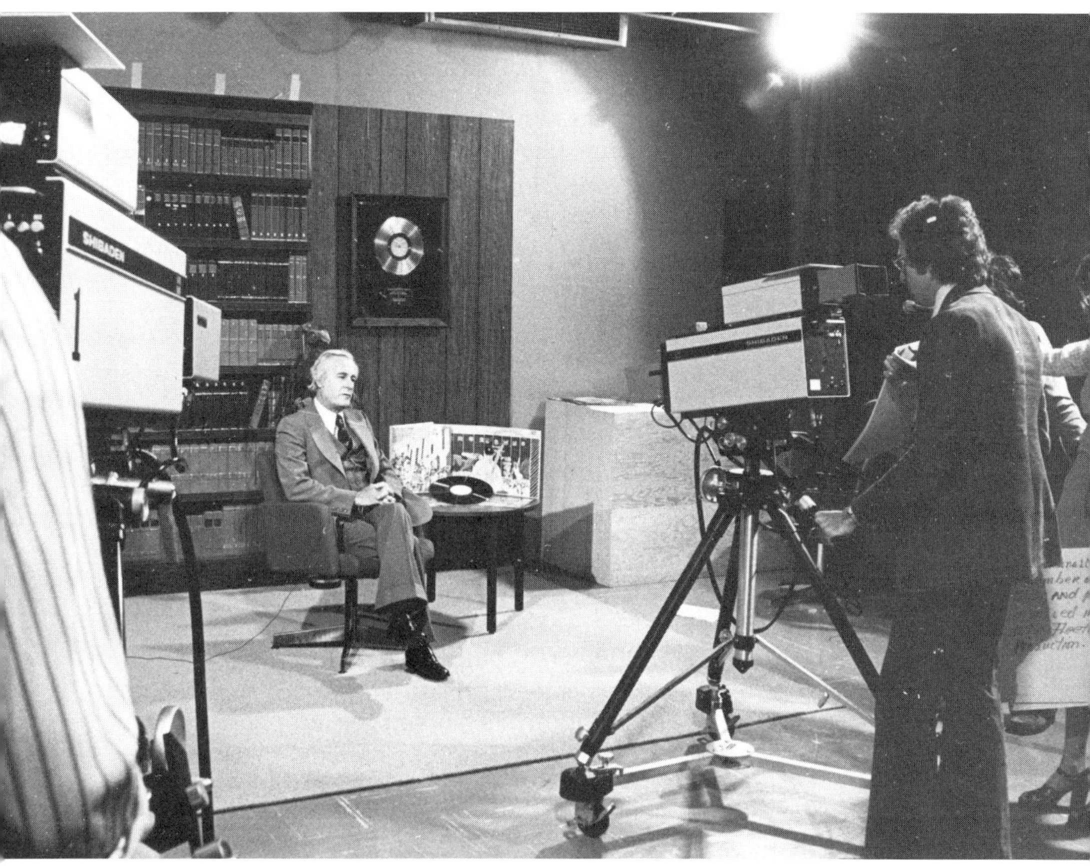

(Photograph courtesy New England Video Services)

Small industrial and educational television studios can be rented for many types of promotional and instructional video tape productions.

number of times. The production house may charge you a tape rental fee for the 2-inch master recording, and then turn the tape over to you until you have had an opportunity to get a second-generation 1-inch master from which you will make your duplicates.

In some instances, a television production house is set up to duplicate directly to a 1-inch VTR, thus reducing your costs. This will rarely be the case at a local television station. You must be certain that you have access to a facility that can perform the transfer from the 2-inch to the 1-inch format. The standard charge for such service is about $150 an hour, not including the cost of the 1-inch tape.

There is usually no loss of quality in transferring from the larger to the smaller format. In fact, in some cases, the quality may be enhanced, in much the same way that a reduction from 35mm motion picture film to 16mm film results in sharper images. If there is a loss in quality, the trouble may lie with the personnel or equipment performing the transfer.

A minor cost item, but one that you should consider, is what is termed "protection usage." Basically, this means renting a back-up recording of your show. For an average cost of $50 an hour, the production house will record the show on two VTRs — the second recording merely as a precaution against electronic problems showing up on the master recording. Once the master recording is verified free from interference or noise, the back-up tape is erased.

Having once contracted with a production unit, it is advisable to let the professionals run the show. You should merely state your objectives; give some indication as to the types of materials you want to include; and who you have in mind to appear on camera. The more locked-in you are to a preconceived format, the less opportunity you offer for creative thought from those you have hired. It is the director's responsibility to determine visual effects, stage arrangement and movement, camera angles, etc. The director wants your suggestions but also wants a free hand in rejecting them if he or she so desires. And if you insist upon making a number of changes — especially at the last minute — you must be prepared to pay for them in both quality and dollars.

If this discussion on the use of outside facilities has appeared to emphasize costs, and possibly sounds discouraging, it is because we wish you to avoid surprises when you receive your production bill. And it is quite possible to negotiate a show for a flat fee that includes all the incidental costs we have mentioned. If you know what you want to do and are prepared to keep to the script, most outside studios will offer the total production for a single price. They ordinarily, however, use the detailed cost method because producers rarely stick to what they say they want, and extra people and overtime must be accounted for.

It is possible to realize still further cost reductions if you contemplate renting the facilities periodically using a standard format. The technical aspects will have been worked out and rehearsal time can be kept to a minimum. It may even be possible to do two or three shows at one time by keeping the crew for an extended period and scheduling the talent at appropriate times.

There are many advantages to using outside production facilities, not all of which are considerations of costs. There is also the matter of technical capability and production quality. The independent, outside facilities must be maintained technologically up to date if they hope to stay in business. They must purchase the latest equipment and staff themselves with skilled personnel to be certain of getting the most out of their investments. This is a luxury most small-studio owners cannot afford. In any case, whether your reasons are for cost or quality, it's worth investigating outside production sources.

The following is reprinted from the *Rate Guide and General Information* card issued by an East Coast educational TV station.

RATE GUIDE – GENERAL INFORMATION

1. All rates included herein are net.
2. Locations not covered by these rates will be quoted upon request.
3. All materials, designs, sketches and floor-plans supplied, constructed or purchased by the station remain the property of the station, unless otherwise agreed to by the station, in writing, prior to show presentation.
4. The usage of all personnel must be in accordance with current labor union contracts and operating conditions.
5. The furnishing of facilities, services and materials covered by this Guide is not guaranteed. All orders for them will be subject to availability.
6. All orders for production facilities and services that cause the station additional expense due to their lateness will be subject to an additional charge as compensation for these costs.
7. All rates are subject to city and state taxes where applicable.
8. All data herein are subject to change without notice.

STUDIO USAGE

The rate applies to the studio's regular lighting and technical equipment. All items ordered over and above these facilities will be charged separately. The rate for studio usage begins with engineering set-up time and ends with tape or air. Charges will be prorated to the next half hour. There is a two-hour minimum.

Studio (color) with 4 cameras: $350 per hour
Studio (color) with 3 cameras: $300 per hour

VIDEOTAPE FACILITIES

The rate includes manpower and is applicable to recording, playback, physical editing, viewing and dubbing. A minimum of two machines is required for recording. Charges will be based on total elapsed time used or ordered, whichever is greater, and will be prorated to the next half hour.

$100 per machine per hour

VIDEOTAPE STOCK

Rate includes new and fully evaluated stock, reels, and shipping cases if needed.

$5 per minute

PERSONNEL

All hourly rates will be prorated to the next half hour, except stagehands and lighting director rates which are prorated to the next full hour. There is a two-hour minimum.

Associate director, stage manager, stage hands: $14 per hour
Engineers: $18.50 per hour
Lighting director: $16 per hour
Graphic artists: $15 per hour
Make-up: $12 per hour
Director and scenic designer: $200 per day

REMOTE FACILITIES

Contains four cameras, and may be used with an auxiliary video tape truck. Rates will be charged from the beginning of engineering set-up to the end of engineering breakdown. A ten hour daily minimum will apply. All manpower used will be charged at the appropriate rates.

$1200 per ten-hour day
$500 per travel day plus 50 cents per mile

FILM/SLIDE CHAINS

$75 per hour

FILM EDITING

Cutting room: $40 per day
Editor: $17 per hour

FILM ANIMATION

Manpower and stand: $30 per hour

AUDIO TAPE EDITING

Manpower and console: $25 per hour

REAR SCREEN

$90 per day

TELEPROMPTERS

$25 per day. Four-hour minimum

CHAPTER 17

POSTPRODUCTION EFFORT

EDITING

By the very nature of the electronics involved, any postproduction editing is difficult and may lessen the quality of the final product. Unlike motion pictures — where film laboratory work can correct certain errors or add dramatic effects without diminishing the picture quality — video tape can receive only limited altering before becoming unviewable.

We recognize that even a great deal of careful planning cannot totally eliminate technical bugs, talent errors, production errors, and so forth. But these can be corrected using proper postproduction techniques.

With the use of sophisticated editing and processing equipment, a good second-generation tape can be obtained and used for duplication purposes. However, if the changes are extensive, the best results may have to come from an outside production house that is equipped with extra stabilizing equipment such as processing amplifiers and time base correctors.

It is advisable to try to get all the audio and visual material on the original master recorded tape. That is, plan the production and cues so carefully that film and slide inserts can be transferred to the original tape during the running of the show. The same principle applies to audio inserts such as music, off-camera narrations, and sound effects.

The term *generation* refers to the number of times successive duplicates are made from an original video tape. The original (or master) material is referred to as *first* generation, while a dub (duplicate) of it is known as *second* generation, and so forth. With each generation the quality is reduced.

Postproduction effort, at best, should be confined to a single edited version of the show (or second generation tape). A third generation tape is possible but with ultimate loss of picture quality. A fourth generation tape is stretching one's luck. (A 1/2-inch duplicate of a 1-inch edited master tape is a third generation tape.)

To understand the basic principles of editing, one first of all must think in terms of picture sequences. For example, if a person on camera turns from the camera to focus attention on another person (or object), at some point the audience will have a natural desire to also look at what the speaker is looking at. If the speaker is addressing an auditorium audience, then at certain intervals an audience shot can appear on the screen with the speaker's voice continuing during the sequence.

If, in this same example, you wish to eliminate a portion of the speech and the resulting deletion would break the continuity of movement (the final words show the face turned left and at the point you want to pick it up the face is turned right) then a cut-away shot should be inserted before returning to focus attention on the speaker. If this cannot be achieved during the actual taping, then the situation calls for editing — electronic editing.

To understand electronic editing, we must first reiterate that the video image and audio sound is the result of a magnetic field laid across a selected portion of an oxide tape. If that tape is cut, the magnetic field is broken and unless the ends of one tape are spliced *exactly*, the recording heads will not pick up the signal. Thus, the tapes are not cut, but are instead electronically edited. (Motion picture film can be spliced because sprocket holes can be matched; audio tape can be spliced because of less critical magnetic fields.)

Electronic editing basically means that one erases one magnetic field and replaces it with another. In its simplest explanation, you can erase both the audio and video portions of a magnetic tape and rerecord another audio and video insert. Or, you can rerecord *either* the audio or video portion, while retaining the other.

In the case of *audio only* editing, the procedure is relatively easy, and not unlike the editing of a simple audio tape. You can erase the desired portion (a word, a sentence, etc.) and insert what words you wish — timed exactly to fit or timed at less than the original with the remaining portion left blank. If the camera is not on the speaker at the time of the required insert, then no changes in the video are required. If there is a problem of lip-sync, then an image edit is required.

In the case of video editing, the critical area is the laying down of a new magnetic field over an old field. Our experience has shown that to rerecord a new image as the machine is erasing the old image is sometimes not satisfactory. The tape guides on the machines are not accurate enough to line up the tape *exactly* in the same position

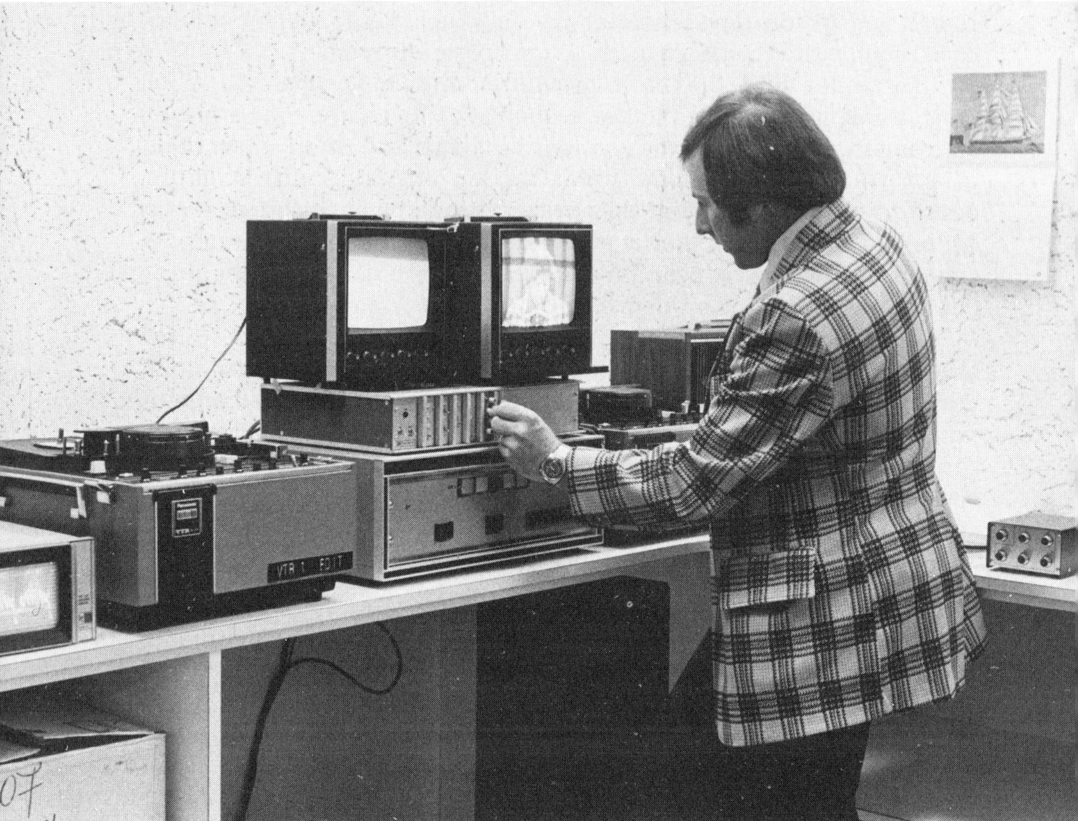

(Photograph courtesy Chase Manhattan Bank, N.A.)

Videotape editing system that consists of two Panasonic 1/2-inch VTRs, backspacer, processing amplifier, two TV monitors, and a waveform monitor

of replay as the original tape recording. Thus, the slight difference may result in *two* images on the same tape. It is better to completely erase the old video image and then return and insert the new magnetic field.

There are two basic types of video edits: insert and assemble.

An insert edit means literally to insert a new picture (and possibly a new audio track) on a tape in a precise amount of time between two existing pictures. For example, a lecturer refers to a piece of machinery which was too large to be brought to a studio or cannot be displayed adequately with a still photograph. You may wish to insert a video tape shot of the equipment in the middle of the lecturer's remarks; the narrator continues; and then you return to the in-studio shot of the lecturer continuing with the discussion. Getting this video shot of the equipment on the screen (and the equipment can be shot before or after the studio production) and in sequence at a precise word or pause and then off at an exact moment is an insert edit.

An assemble edit means you wish to take a number of different images and narrations and tie them together for a proper sequence. This is the type of editing required if you are showing a one-hour video presentation of the highlights of an all-day seminar, with an off-camera voice supplying the transitions between speakers and topics.

It must be cautioned that not all VTRs advertised with editing capability are capable of performing an insert edit, because some VTRs are automatically erasing the tape a few seconds ahead of a new recording and are simultaneously laying down a new control track — the magnetic track that adjusts the electronics in the VTR so that the video and audio heads are properly aligned to read their respective tracks. An assemble edit, by virtue of its being a new sequence of shots, automatically requires a new control track as it progresses, and all VTRs perform this function.

Mechanical splicing of video tape for editing purposes is possible, but is normally an unacceptable method of editing in a helical scan recorder. Since the video tracks are laid down at roughly a 5° angle, in order to achieve a frame edit, it would be necessary to cut the tape from one side to another at this angle, keeping the cut to within an accuracy of 0.005 inch. No equipment has yet been devised to do this accurately. If a normally vertical or slightly slanted cut is made, even if the control track is perfectly aligned, there will be a vertical wipe up the screen during each edit. The picture will also have a tendency to break up at this point. Even if a very good splice is made, there is still a tendency for the heads to clog during the passage of the splice through the machine. Splicing should only be done if the tape has been physically damaged.

But to reiterate, the technique is not advised, and should only be used if electronic editing is not possible.

Most viewers have come to expect a *sound* to accompany a *picture* when viewing a television show or movie. The lack of sound is, itself, a *sound effect* — usually meaning that the silence will be shattered dramatically. Therefore, even the opening or closing of a single industrial or educational show should have some type of sound as background for the title cards or establishing shots. In most cases, this sound is made by music, but the opportunity to utilize actual sounds or sound effects should also not be overlooked. For example, the sound of electronic testing equipment or a recording of a data-processing room can be the sound-signature to introduce a number of industrial tapes.

When using music, avoid using music that is either popular or identifiable (except in the case where the music is specifically identified with the company or school). The audience will have a tendency to "sing along" with the tune, or put in the lyrics, or make other identifications. Libraries of mood music are easily obtainable and not expensive.

In most cases, the music or sound recording should fade in while the screen is black and just prior to fading in the opening title card. The sound is then lowered or kept "under" as the narrator gets the show under way. Once the main speaker or topic is introduced, the music or sound can fade out completely. There are no hard or fast rules on this; a director must sense when a background sound is required and when it is distracting.

The same situation applies during the course of a show and when the show ends. A series of filmed factory shots can be enhanced by music that has a mood of whirring energy. Or the shot of a city street can be more effective if one can hear the traffic noises — either up full or as background to an off-camera narrator.

When ending the show, the sound can linger a few seconds after the last fade-out, and then the sound also fades.

TRANSFERRING AND DUPLICATING

There are two methods for transferring and duplicating video tapes. The first is a head-to-head method that involves a master machine feeding slave machines. The other is a high-speed transferring process which involves contact duplication, but which is not yet generally available.

For limited numbers of copies, a master/slave system can be part of a small-studio operation. High-speed duplicating machines are extremely expensive and are best utilized from firms which are in that business.

In the in-house facility a master machine can feed a single slave machine (either 1-inch or 1/2-inch) with an unamplified signal and obtain quite satisfactory copies. If more than one slave is connected to the master recorder, then a distribution amplifier should be used. The distribution amplifier will provide equal, isolated feeds to each

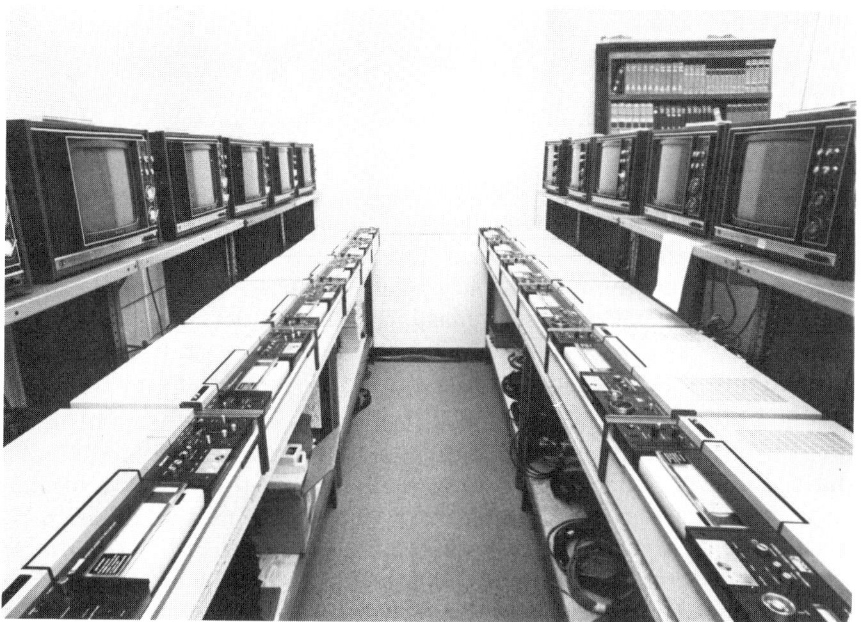

(Photograph courtesy New England Video Services)

Video cassette duplication facility using standard player/recorders with distribution system from single source

machine. It is important, in video, that each video line be accurately terminated and in feeding a number of machines it is considered the best practice to feed each machine separately and to provide an accurate termination at the machine.

When duplicating from one machine to another, it is advisable to seek compatibility through brand and model. If different makes are used, it is best to dub from the wider format to the smaller, and from the higher quality recorder to the lesser one. In both instances — format and quality of product — the main factor is the stability of the picture, which cannot hold as well going up in size as it does being reduced.

The major problem with head-to-head duplication is that it is a relatively slow process and involves a substantial amount of maintenance time. A tape is transferred at the same rate of speed it was recorded — thus a one-hour tape takes one hour to transfer. In addition, for each transfer, the video heads of the master VTR and the slave machines should be cleaned.

Recent advances in contact, high-speed duplication and better quality tapes have made this system of transfer more practical than head-to-head transferring.

There are two basic systems for the contact duplication of video magnetic tape: a magnetic process or a thermal process.

Magnetic Process. This is a system in which the master recorded tape and a blank tape are brought face to face so that their oxide layers are pressed together. The two tapes are then passed through a copying station where a magnetic field is applied to them (the field transferring the pattern of the master to the copy).

As the two tapes come out of the copying station they are wound onto their respective spools. Because the master tape has been contact-printed, the copy is a mirror image of the master.

In order for the copy to play on a conventional machine, it must have a conventional track pattern. That being the case, the *master* tape must have been recorded so that *it* has a mirror-image track pattern, and this requires a specially adapted recorder.

The difference in signal degradation between the master and the copy is on the order of about 5 db in the signal-to-noise ratio, and the process shouldn't appreciably impair the quality of the master tape. Usually, master tapes are made on chromium-dioxide tape, which has some advantages over conventional ferric-oxide tape.

A feature of this process is that the longitudinal tracks on the master tape (audio and control tracks) can only be dubbed from the master to the copy via high-speed dubbing methods (*after* the contact duplication of the video tracks is completed).

Thermal Process. In this system, the recorded master tape and the copy tape are brought face-to-face with their oxide layers pressed together. They are then passed through a copying station where they are heated to a certain point (the Curie point of the copy tape). The tapes are then cooled and spooled on their separate spools.

When brought to its Curie point, the copy tape changes from a ferromagnetic to a paramagnetic state and becomes completely demagnetized. When it returns to ambient temperature, it is then ferromagnetic. At the point of passing between the two points it is highly susceptible to the magnetic orientation of the master tape and the copy is thereby made.

There is little degradation of the master tape in this type of copying operation, and 100 copies can be made without appreciable deterioration. Another important feature is the fact that the audio and control tracks can also be transferred directly, eliminating the need for a separate dubbing operation.

The copying speed of this system is quite high.

ELECTRONIC FILM TRANSFER

In addition to magnetic tape transfers, it may also be desirable at times to make a motion picture copy of a video tape program since film is to date the only international standard color medium. Known as an electronic film transfer, or tape-to-film transfer, it permits distribution of a program where VTR equipment is not installed.

In tape-to-film transfer, a film camera records the picture directly from a special screen, which in turn is being fed a picture from a video tape recorder. The processed recording is 16mm film and can be played on standard projectors of that size.

You can expect some loss of picture and sound quality from a tape-to-film transfer.

EQUIPMENT MAINTENANCE

A major part of any postproduction effort must be the proper maintenance of the equipment so that it will perform as expected.

The equipment that must be maintained consists of both the VTR upon which the show is recorded, and the units upon which it will be played back. Lack of maintenance at either end of the production cycle can completely destroy days or weeks of planning and shooting.

The following maintenance tips are reprinted from *Producers Manual* produced and distributed by Magnetic Products Division of 3M Company.

VTR Maintenance

The combination of a well-adjusted and maintained video recorder and a high-quality magnetic tape will assure the production team of excellent results in faithfully capturing everything that is being seen on the monitor. However, if either the tape is defective or the recorder is not functioning properly, a degraded picture or even no picture at all will be seen on playback.

Recording tape itself can be considered a passive device since the operator can do nothing to change its basic characteristics. And "adjustments" to ensure peak performance were predetermined and became part of the basic formula for the oxide coating. Variables affecting picture quality and consistency can most often be attributed to the external or internal controls on the recorder itself. This, of course, is not to say that problems cannot arise because of the recording tape, for if a tape of inferior quality is used, the finest recorder made will not be able to record a quality picture on it.

One of the more common symptoms that is observed by recorder operators when experiencing difficulty is described as noisy picture. This is also sometimes referred to as "grainy" or "snowy." There are several causes for this and any one of them alone or in combination can disrupt the picture quality.

As an aid to the operator, we have listed those areas that should be checked if noise is noted when the tape is played back. We, of course, have assumed that the signal being fed from the program line to the recorder is normal and the recorder heads and guides have been properly cleaned.

Inadequate Head-To-Tape Contact. During both record and playback, transfer of information between the head and the tape depends upon their intimate contact. The video head is said to actually penetrate the tape surface by its protrusion beyond the drum assembly. The picture noise will increase as the amount of contact decreases. On those recorders with two video heads, one may be making proper contact while the other is not. In this case alternate fields will be affected. Because the speed of 60 fields per second is faster than the eye can resolve, the combined effect of good and noisy fields blends into a picture that can range from moderate to severe noise. On single-head recorders, the noise can be even more pronounced even to the point where on playback, only noise is seen and the picture information is totally lost.

This is not to be confused with mistracking, which causes a horizontal noise bar in one portion of the picture only. With less than adequate head-to-tape contact, the entire picture is affected and most often is completely obliterated by noise. If loss of head-to-tape contact is suspected, the amount of head protrusion beyond the drum should be checked by a qualified serviceperson to see if the head height is within the recommendations of the recorder manufacturer.

Feed Reel Holdback Tension — Or Take-up Reel Tension. The tension between the reel and the capstan holds the tape in contact with the video head. If the tension is less than adequate, symptoms like those described above can be observed. While the tension is sometimes a front panel operator adjustment, it is often an internal adjustment that must be set up by a serviceperson. There are some cases where a misadjusted tension control will cause symptoms that look like mistracking.

Improper Video Head Drive Current. In the video recorder, the incoming picture information is used to modulate a radio frequency carrier. This modulated R.F. is then fed to the video heads. If the level of R.F. drive current to the head is improper, not enough signal will be recorded on the tape. On playback the signal-to-noise ratio will be poorer than desired. The result will be a noisy picture. The drive current adjustment is an internal control and should be adjusted only by a qualified serviceperson who has the necessary skill and equipment. It is *not* the "video gain" or "video level" control found on the front panel. We recommend that the drive current be checked periodically or in accordance with the recorder manufacturer's suggested procedure. It must be set up when new heads are installed and should be checked and adjusted at least once during their life, preferably when they are worn about halfway. Checking at closer intervals is advisable for overall picture excellence.

Video Head Is Worn Out. When the video head is worn beyond its useful life, the problem usually first manifests itself as difficulty in recording. The head may still play back tapes that have been recorded at an earlier date or on another machine. Not only will the head protrusion be below normal but the gap between the head pole pieces will have widened, changing the head characteristics. The development of symptoms would fall in this order.

1. Recently made recordings look noisy on playback. Old recordings play back OK.
2. Impossible to record on machine. Playback of all tapes is noisy.
3. Impossible to record on machine. Impossible to play back all tapes. Noise is severe.

If the video head is worn out, adjustment of either drive current or head protrusion will be of no help and the head or heads must be replaced.

Poor Slip-Ring Contact. The electrical energy sent to the head during recording, or taken from the head during playback, must pass through a slip-ring connection or a rotary transformer. This is necessary because the head assembly is rotating and the recorder is stationary.

A slip-ring can be described as a stationary arm or "brush" that rides against a smooth ring or disc that is attached to the moving head assembly. If the point of contact between the brush and the ring becomes dirty, there will be momentary losses of picture information. These will appear on the playback monitor as a series of bright white flashes.

The slip-ring assembly can be easily cleaned by using a cotton swab moistened with Freon TF (a DuPont trademark) or Genesolve-D (an Allied Chemical trademark). This should only be done when the video head assembly is not turning, as damage could result if the swab became entangled. If cleaning does not eliminate the problem, it is possible that there is not enough pressure being exerted by the brush against the ring. If this is found to be the case, the unit should be adjusted.

Some recorders are equipped with replaceable carbon brushes that will wear down in time. If inspection discloses that they have become worn, they should be replaced.

On those recorders using rotary transformers, no adjustment is usually possible and if the transformer malfunctions, it must be replaced. These are rugged units, however, and seldom cause trouble.

Faulty Record and Playback Amplifiers. The amplifier used to process the signal for recording is different from the amplifier used when the tape is played back. For this reason difficulty may be encountered in only one of the modes of operation and the other may function normally.

If previously recorded tapes play back normally, but a tape that has just been recorded cannot be adequately played back, one might suspect the record amplifier. On the other hand if there is difficulty in playing back all tapes, the trouble could be in the playback amplifier. The symptoms can range from minor picture degradation through severe noise to complete loss of picture.

Inadequate Limiting. In the playback process, the video signal undergoes amplitude limiting to "clean it up" on the way to the demodulator where the intelligence will be removed from the R.F. carrier and ultimately sent to the monitor. If the limiter section is not doing the proper job, the inherent system noise will be passed unchecked. The result, of course, will be a noisy picture.

Because of the internal design of the helical recorder, when in the record mode, the picture seen on the monitor has been sent through a great deal of the record and playback electronics. If the limiting is inadequate, this will be seen on the monitor as a noisy picture possibly while recording is taking place, as well as when the tape is being played back. However, the noise will nearly always be more pronounced during playback.

Equalization Out of Adjustment. The playback electronics of the recorder must pass the range of frequencies that are needed to reconstruct the picture that was recorded. A control within this circuitry, known as the equalization control, adjusts the response of the playback electronics to adequately pass the necessary band of frequencies.

If the equalization control is not properly adjusted, the picture, observed on the monitor during playback, can display black streaks in the areas showing the most white. The adjustment of the equalization control should be performed only by a qualified serviceperson.

FM Carrier Off Frequency. As mentioned earlier in this chapter, the video information to be recorded is not sent directly to the video head, but is used instead to frequency-modulate an R.F. carrier. The resultant, modulated R.F. is what is then recorded.

If the frequency of the carrier is higher or lower than it should be, the signal will not match the other tuned circuits in the record and playback system. If the carrier is off frequency, the picture, when played back, will exhibit a very odd appearance. Certain portions of the picture will be degraded by interference of a moire or herringbone line pattern. This type of interference, while broadly classed as noise, appears different from the more typical "snowy" noise referred to under the other headings of the section. The more the carrier is off frequency, the more severe will be the picture degradation.

Because of the recorder design, the symptoms of a severely off frequency carrier will be seen on the monitor during record as well as playback. This is a delicate, internal adjustment and should not be touched by anyone except a trained serviceperson, equipped with the proper instruments.

Video Gain Too High. The front panel "video gain" or "video level" control adjusts the amount of picture information that is entering the recorder. If this is set too high, too much video information is being modulated onto the R.F. carrier. The result, when the tape is played back, is seen in the brightest white areas of the picture. These white areas will actually tend to appear darker than expected and the picture will "look negative."

This problem can be corrected by reducing the setting of the video gain control. Of course, if the control setting is reduced too much, the picture, when played back, will be weak and lack contrast. Always follow the recommendations of the recorder manufacturer when adjusting the video gain.

External Electrical Interference. Sometimes fluorescent lights, large motors, faulty appliances, or dirty switches in the area of the studio and control room will cause interference to be recorded onto the tape. Unless the interference is quite pronounced, it will probably not be seen on the monitor during recording, but will be seen when the tape is played back.

This type of disturbance manifests itself as white flashes in the picture. These white flashes can be as small as specks or might be white horizontal lines as long as 1/10 the picture width. They may tend to be grouped in bands or can occur in completely random fashion. If this symptom is seen, the source of the interference must be located, and corrective action should be taken to repair it.

Noisy or Weak Signal to the Recorder. It goes without saying, that the recorder can record and play back a picture only as good as the one it originally received from the program line. If the video to be recorded was weak, distorted, or noisy, the picture, upon playback, will be an accurate reproduction and include all the faults of the original.

If the recorded signal was too weak, the normal tendency upon playback is to increase the monitor contrast. While this will help to darken the blacks and lighten the whites it will also make the system noise more noticeable. It is important, therefore, that the best possible picture be fed to the recorder so that a high-quality recording can be made. If this is done, and the recorder has been properly adjusted and maintained, the quality of the picture on playback should be excellent. It is, of course, important to use a video tape that has been designed for use with the particular recorder, and is itself a high quality product from a reliable manufacturer.

Playback Units

Generally speaking, those recorders using the same tape width and built by the same manufacturer do offer the capability of tape interchange. If both the machine upon which the recording was made and the machine upon which playback will take place are adjusted according to the manufacturer's specifications, successful interchange should be achieved.

If difficulty is encountered in an attempted interchange situation, possibly one or a combination of the following impediments are responsible.

Supply Reel and Take-up Reel Tension. If either the machine doing the recording or the machine playing back the tape is set up so that reel tension is not within the manufacturer's specification, interchange may be impossible. If this is the case, even adjustment of the front panel tension control (if the recorders have one) will probably not be of any help.

This tension difficulty may exist on either recorder. If the unit doing the recording is at fault, it may replay its own recordings without problems because the tape is being subjected to the same tension conditions in both record and playback. This machine will not be able to play back tapes made on other recorders, and other recorders will not successfully play back tapes made on the machine with the problem. It is most important that the unit doing the recording be accurately adjusted. The reason for this statement is obvious. If the machine doing the recording is improperly adjusted, there is the danger that operators of the machines used for playback, may adjust them to accept the tape from the maladjusted recorder. This would result in several machines operating in a nonstandard way.

Sticking of Rotary Idlers. Many recorders are equipped with rotating idlers that guide the tape into and away from the head drum assembly. If these idlers fail to rotate smoothly, tape travel will be impaired. This can cause trouble when tapes are interchanged, and can even affect the playback of tapes recorded on the same machine. Anything that affects the smooth and free travel of the tape in its path from supply to take-up reel can cause difficulty.

Capstan Becoming Polished. The capstan controls the speed of the tape traveling through the recorder. On most helical recorders, it contacts the smooth-coating side of the tape that is held against it by the pressure of the pinch roller on the backing side of the tape. As the capstan turns, the tape is moved. If, after many hours of use, the normally machined surface of the capstan becomes polished, it may tend to slip against the smooth tape. If this happens, the tape will not be moved through the recorder in a steady fashion. If slippage is severe, tape speed will be decreased. Replacement of the capstan is recommended if it appears to have become too smooth from use.

Pinch Roller Pressure. The pressure of the pinch roller holds the tape against the capstan so that the rotating capstan can move the tape at the prescribed speed. If the pinch roller is not exerting the proper pressure against the tape to force its contact against the capstan, this will result in the same symptoms as a polished capstan mentioned in the preceding point.

The amount of pinch roller pressure is specified by the manufacturer for each type of recorder and should be adhered to. This adjustment should be made by a qualified serviceperson.

Angle of Tape Path through Video Head Assembly. The guides that control the tape entering and leaving the video head assembly are adjusted to hold the tape against the rotating head at the proper angle. These guides actually determine the angle of the recorded track. If they become damaged or misaligned, the tape will be guided at the wrong angle and the recorded video track angle will be improper.

This will not normally affect tapes played back on the same machine, because the playback angle and record angle will agree. It does, however, become critically important when tapes are interchanged. As the record and playback angles differ, the symptom could be a noisy picture if the angle difference is slight, or complete mistracking if angle difference is more severe.

Worn Stationary Guides. Many recorders are equipped with stationary guides that control the tape position. These stationary guides are designed so that the tape passes over a smooth, curved, central portion and is held in place by flanges at the upper and lower edge. After prolonged use, the flanges of these guides can become grooved because of their constant contact with the edges of the moving tape. As the groove develops, the tape can begin to wander up and down between the flanges. This will affect the angle of the recorded track and display symptoms much like those described in the preceding point. Worn guides, such as just described, will also affect playback of tape recorded on the same machine, as the tape will change its angle in a random fashion.

As the tape passes these worn guides, the tape edges can become damaged by their contact with the grooves. The tape's coating as well as the backing can be worn, and bits of debris can be deposited on the surface of the tape. This will appear as dropouts when the tape is replayed on any machine. Stationary guides should be replaced as soon as wear is noticed. Pay particular attention to any buildup of oxide being scraped from the tape and deposited in the area of the guides.

Defective Control Track Head. During recording, the control track head records a series of pulses along one edge of the tape. On playback, the same head "reads" these pulses and uses the information to control the speed of the rotating head so that accurate tracking will be maintained. These control track pulses are to tape as sprocket holes are to motion picture film. In fact they are sometimes referred to as "electronic sprocket holes." If the control track head is worn or has been damaged on the machine used to make the recording, the recorded tapes cannot be played back on other machines. If the machine used for playback has this problem, it may not be able to play tapes from any machine.

A defective control track head will also affect tapes that are recorded and played back on the same recorder. If a defective control track head is suspected, it should be checked, and replaced if necessary by a trained serviceperson.

Before leaving our discussion of the control track head, we must also mention that loss of contact between the head and the tape will produce similar tracking errors even if the head itself is functioning normally. If the mounting assembly has caused the control track head to be moved out of position and away from the tape, the necessary intimate contact between tape and head will be disrupted. The audio head can also suffer from mispositioning and will either record or play back a poor sound track or even no sound at all. Head position is important and should be checked periodically.

Capstan or Head Drum Servo not Functioning Properly. The speed of the rotating head during playback, and sometimes the speed of the tape itself, is closely controlled by electronically monitoring the control track pulses. If the electronic servo control circuits should malfunction, a degraded playback will result. The symptoms could range from a slightly noisy picture to complete loss of stability and total mistracking. These servo circuits should be checked, adjusted, or repaired only by a well-trained serviceperson who possesses the necessary skill and equipment.

In this chapter, we have attempted to enumerate those items that must be considered by recorder operators, when recording, playing back, duplicating, or interchanging video tapes. We strongly recommend that recorder operators become as familiar as possible with the operation of their equipment and with the basic principles of video recording.

Television recorders should be kept clean and well adjusted at all times. If even a slight symptom of abnormal operation is noticed, it should immediately be checked and the necessary repairs or adjustments should be completed before the machine is returned to service. This will result in the machine being ready for service whenever needed, will actually reduce the total amount of money spent for service, and will ensure that recorded pictures and sound will be of the highest quality.

CHAPTER 18

STORAGE, HANDLING, AND DISTRIBUTION OF TAPES

It is important to treat video tape carefully, since proper handling will enable the continued reuse of the tape. If handled improperly, the effective useful life of the tape will be shortened considerably. Damaged or dirty tape can also contribute to damage to the video recorder head and other parts of the mechanism.

How long a single tape will last depends on (1) handling and storage, (2) re-records, and (3) number of playbacks. The last factor — playbacks — is a relatively minor item. There seem to be no valid tests indicating the life of a tape that is properly handled and merely played back on a well-maintained unit. One tape manufacturer ran a test on two quadraplex tapes for 4400 playback passes, and the results showed no tape wear or dropout.

The second factor — re-records — is a matter of how well the tape is erased. Our experience has been that if a tape is bulk-erased, one can expect five to seven programs to be successfully transferred before "cross-talk" begins to interfere and cause double images and severe picture dropout.

Naturally, poor physical handling of video tape can cause creases, dust, stretching, and so forth. Once a tape has been mishandled, there is little that can be done to salvage it.

The tape on reels constitutes a considerable weight; thus care must be exercised in the loading and unloading operations. Otherwise the heavy reel can fall on the head or tape transport mechanisms of the video recorder.

The reel should be handled with both hands when the recorder is being loaded or unloaded. It should never be lifted by the top flange of the reel, but should be carried by the hub. The flanges should not be squeezed, because this can cause damage to the edges of the tape.

If a reel of tape is dropped, it is almost certain that either the reel or the tape itself will be damaged to some extent. There is no way to repair creased, scraped, or bent tape.

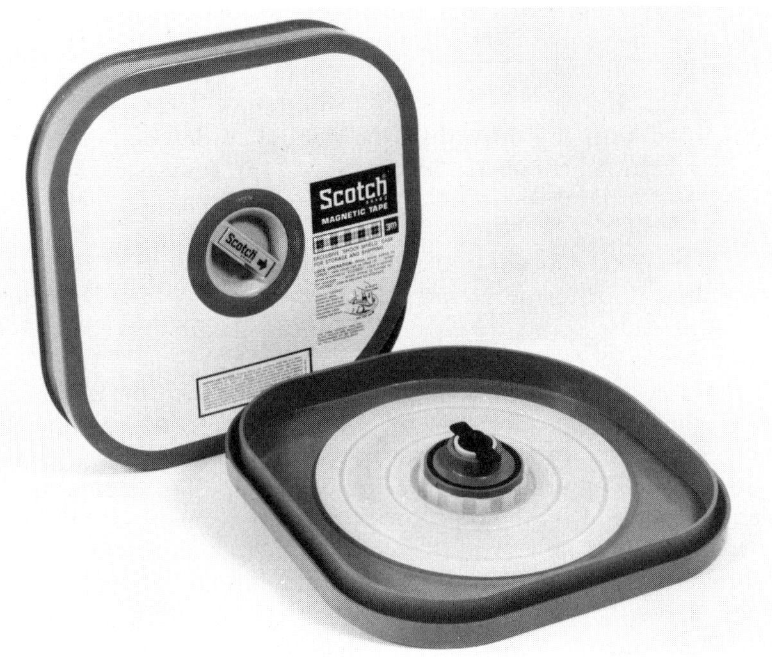

(Photograph courtesy 3M Company)

Special case protects video tape from handling, storage, and shipping accidents

When tape is to be stored it should be packed in the box provided by the manufacturer. The reel should also be placed in a plastic bag (also provided by the manufacturer) before being put in the container. This will offer further protection against dust. As mentioned earlier, dust and dirt constitute a severe problem in video tape production, since they can cause disturbances in the picture as well as damage to record and playback machines.

Generally speaking, it is the best policy to store video tape on edge, in the same way that books are stacked on a shelf. If boxes of tape are stored lying flat there is the danger that several tapes may be stacked on top of each other, and there is always the chance that the reel on the bottom will be bent by the weight of the other reels.

We have mentioned earlier that the studio and directing booths needed to be kept scrupulously clean, and controlled, if possible, in terms of temperature, humidity, and air filtration. The storage area for tape should be as carefully controlled. It is generally agreed that a temperature of about 70°F. and a humidity of 50% are ideal for the storage of video tape. Extremes of temperature or humidity can cause serious problems and should be avoided. (Tapes should not, therefore, be stored near windows where direct sunlight can fall on them.)

Persons responsible for storing and using video tape must be well acquainted with the operation of a video recorder and how properly to thread the machine. If the tape is not carefully placed in the tape guides and around the drum heads, scratches, tears, and edge damage will result.

Also, if tape is allowed to become dirty (because it is dropped on a dirty floor, or unwound on a dirty work surface), the dirt and dust will cause scratches along the entire length of the tape. Once this happens there is no remedy for it.

It is also important to ensure that proper tension is maintained when video tape is rewound on the supply reel. At no time should tape be cinched on the reel (for instance, by trying to tighten up the tape on a reel by tugging on the end of the tape), since this will cause cinch marks and scratching. For the same reason, a reel when being rewound should be allowed to coast to a stop on the machine. Too abrupt a stop can cause the tape to slide around the hub and make cinch marks.

Since video tape is in fact magnetic tape, care should be taken so that it isn't stored near any powerful source of electromagnetic energy — such as X-rays, or radioactive sources.

Should some problem appear with video tape, the trouble is usually with the recorder or the playback unit. Seldom is the trouble attributable to the tape itself.

Tips on tape handling could be summarized as follows:

1. Do not drop a reel of video tape. This will bend the reel flanges, causing direct damage to the tape and also causing the tape to feed improperly on the recorder.

2. Badly wrinkled or damaged tape ends should be cut off cleanly. Trying to thread leaders of this type on the recorder can result in misalignment of the tape, with possible damage to the entire reel of tape, and also to the video head of the recorder.

3. Wherever possible, avoid splicing video tape. If it must be done, be sure to use the proper splicing tape, and use the techniques that will produce splices that won't tend to catch or damage the heads when they run through the equipment.

4. Thread the tape carefully around guides and the head drum. This will help prevent scratches, tears, and stretching which can affect recording or playback.

5. Don't stop the tape in the middle of a reel without releasing the tape tension, or otherwise following the exact instructions pertaining to the equipment.

6. Never attempt to remove tape from the guides, capstan, or head drum while in the middle of a reel of tape. This can cause permanent damage to the surface and edges. The tape should be wound onto one of the reels and then removed.

7. When operating the recorder be sure to use optimum head tip projection. If too much projection is used, the tape is chewed up and its life shortened. If too little projection is used, poor picture quality results.

8. It is absolutely critical that the video tape recorder be kept clean. Otherwise, dirt and oxide deposits can build up on guides, capstan, and heads and cause damage to both the equipment and to tapes.

9. The tape pack should be checked every time it is rewound. Continual rewinding with too little (or too much) tension causes an uneven tape pack, or slippage that can result in extensive tape damage.

An important feature of any tape library or tape-storage facility is some system for identifying taped materials. This is of paramount importance in any production facility, since there is nothing more frustrating than a search through endless rows of boxes, all of which look the same. (Also, unlike film, there is nothing characteristic in the physical appearance of one roll of tape as compared with any other — no picture can be seen on it.

Tape should, therefore, be identified in two ways. There should be a piece of pressure-sensitive tape on the reel of video tape itself, with information on it concerning the name of the production, the date of production, and other pertinent data. The same information should appear on the box supplied by the tape manufacturer.

It should be noted that the pressure-sensitive tape information should not be prepared while the tape is on the reel. The pressure of a ballpoint pen marking on the tape will cause indentations through several layers of the magnetic tape. These indentations can cause poor recording and playback characteristics. The identifying strip should either be prepared on a flat surface, or the information should be typed and the tape applied.

There are any number of library or storage systems that can be employed — but there should always be at least two ways of indicating the content of the tape.

The name of the production (or a description of the contents) can serve as the subject title. The individual reel and box should also be assigned a number. The number can be assigned from a log, with sequential additions with each new production.

The third piece of information that can serve to identify the origin of a piece of programming is the date. Such a dating system can also be a part of a library log. (Further ideas on a filing system are discussed in Chapter 6, "Production Planning.")

Once again, the importance of this activity cannot be minimized. Too much effort and time is expended in the production of a video tape production to risk its loss by sloppy or insufficient storage procedures.

Distribution and inventory control of video tapes is as important as any other factor in the total production. A considerable amount of money can be tied up in tape inventory, and an oversupply (or undersupply) of tape can materially affect total production costs.

Most major manufacturers of video tape will sell and deliver blank tapes directly to the volume user, thus by-passing the distributor or retailer. In addition, one can usually negotiate a further substantial discount for volume purchases, provided he is willing to receive a bulk shipment and do his own storage.

What constitutes a proper inventory, depends, of course, upon the number of productions and whether the productions are freely circulated or are part of a library of tapes. For example, if a company is distributing a weekly tape to 100 outlets which then return the tape to the production facility after viewing, then an on-hand inventory of about 400 tapes is required. That number breaks down as follows:

100 tapes being viewed
100 tapes in transit (coming or going)
100 tapes being duplicated
50 tapes in late return or damaged
50 tapes for backup.

With careful handling and erasing, one can figure on getting approximately seven productions per tape, plus eventually receiving the late tapes. But even so, over a year's time, the producer of a weekly show being viewed at 100 outlets will probably have to purchase over 2000 blank tapes.

Keeping track of that number of tapes is a problem in traffic control, and requires a logging system that must be constantly kept current. There are many ways this system can be maintained, but most distribution centers will want some sort of double-entry system: a record of the individual tape (where it is, what is on it, and how many times it has been dubbed) as well as a card for each user (the tapes he has received and the tapes he has returned). In addition, there should be some way of indicating when tapes are not being returned on time, so that a communication can be sent to remind the balky user not to tie up your inventory.

It is probably sound advice to seek distribution help from anyone in your organization who is familiar with traffic control and mail handling. If no such help is available, it may then be wise to find an outside company to handle your tape distribution. The care and handling of your finished product — like all the prior steps of video tape production — is too important an element to leave to chance.

GLOSSARY

Amplifier An electronic device in which a signal is boosted or strengthened.

Angle shot A shot taken from a position either above or below standard tripod height.

Aspect ratio The ratio of picture width to height, expressed numerically. In television, the aspect ratio is four (wide) by three (high).

Audio Pertaining to sound, particularly to the frequencies heard by the human ear (roughly 20 to 20,000 cycles per second).

B/G or background A term that usually refers to the background of the setting itself, but which can also refer to background music, or background sound effects (either of which are played at a level that is subdued, and below the level of the dialogue of the performers).

Backlighting Light directed on the performer or the subject from behind, so as to create a dimensional effect (offsetting purely front light).

Baffle A movable wall or screen made from spun glass, or hung with drapes, etc., to act as a sound-absorbing surface near a set. The purpose is to prevent echoes.

Balance, lighting The use of the proper proportions of key and fill lighting when illuminating a television studio set.

Balance, sound The correct usage and mix of any elements which blend together in a recorded track (e.g., the use of dialogue, music, special sound effects).

Banding A visible difference in the reproduced characteristics in that portion of the picture associated with one head channel, when compared with adjacent areas associated with other head channels. There may be banding in which the visible difference is in (a) the hue; (b) noise level; or (c) saturation.

Barndoors Panels that fit on the front of spotlights and are adjustable so that the spill of light can be controlled or masked.

Basic set The television setting as it exists without props or other production material on it.

Bayonet lock The spring-tension positioning of a lens which allows its quick attachment to, or detachment from its housing. The lens may be locked into position by twisting.

Beam An electron gun in a camera or picture tube emits a flow of electrons called a beam. The beam creates the scanning lines as it sweeps the tube's target area.

Bear trap A device used in the television studio set, that can be clamped to rails, pipes, posts, etc., and which is used for supporting such items as flags, nets, or small light units. Its spring-clamp feature enables the trap to be used in this way.

Bias The minimization of normal distortion by the introduction of a high-frequency signal during recording.

Black The screen with no video information.

Blank tape Video tape with no recorded signal on it. It may be fresh tape, or bulk-erased tape.

Blanking Relating to the scanning process of a camera or receiver, whereby the retrace lines are blanked out.

Blocking The process used by the director in planning the physical movement of people and objects on the set. Placement of people is sometimes simplified by marking (with chalk or tape) certain prearranged positions on the studio floor. The participant can then "take his or her mark."

Blur An image which is unclear, fuzzy, or in any way indistinct, caused by camera movement, poor camera focus, or instrument vibration during taping.

Board The control board on which are located video (and often audio) controls for cameras, recorder, and microphones.

Boom This term may refer to a camera boom or a microphone boom — the elevating support that suspends it and can move from position to position during taping. To boom a camera up or down means to move the entire camera one way or the other vertically.

Borderlight A striplight which is used on a television set and placed above the acting area.

Break 1. To take time out, as a rest. 2. A director's command to a cameraman, as in "Camera 2, break for a two shot."

Breakup A distortion of the image which occurs momentarily in television or videotape recording.

Brightness The intensity of objects (highlights) on a TV screen.

Brightness control A television receiver control that allows for the adjustment of relative brightness of the screen.

Bulk eraser A device which operates on standard 110-volt AC current and creates a strong magnetic field. It is used to erase TV magnetic tape so that it can be reused.

Business Activity and movement (other than spoken words) to make the performer appear relaxed, lifelike, and believable.

Busy A set is called "busy" when it has too many confusing shapes, details, or other physical elements in it. Such a condition detracts from the performers and their message.

Camera In television production, the electronic device that converts visual images into electrical impulses. It utilizes a standard optical system and a light-sensitive pickup tube.

Camera chain A television camera plus its associated electronic equipment needed to deliver a picture for TV.

Cans The headphone and speaker assemblies worn by cameramen, stage managers, and director during production of a TV show. It makes it possible for them to communicate without their voices being picked up by the microphones.

Capstan A spindle or shaft that rotates against the tape and provides the motion of the tape through a recorder or playback unit. The capstan must be kept clean or a constant speed will not be maintained.

Center This is a director's term that tells a cameraman to frame a subject in the center of the picture. The command may also be to "center up."

Channel The portion of the electromagnetic spectrum (usually 6 Megahertz wide) that is used by a station for the transmission of audio and video information.

Cinching Longitudinal slippage between layers of tape as it is being rewound. It can cause damage to the tape; therefore, take-up pressure must be carefully regulated.

Clipping The process which enhances the videotape picture or signal by keeping the signals above or below certain set frequencies, thus keeping the black and white levels of the picture within set limits. This process involves electric circuits which are built into videotape recorders; they cut off the amplitude peaks of the videotape signals.

Closed circuit A system in which television signals are distributed from originating equipment directly to TV monitors. The distribution system may employ coaxial cable, microwave relays, or telephone lines.

Close-up (CU) A fairly tight shot of the performer or the subject matter. In small-studio parlance, this would refer to the head of a performer. An extreme close-up (ECU) might mean a selected portion of the face, such as the eyes.

Coaxial cable A specially designed cable used to carry one or more channels of TV signals. It makes possible the transmission of high video frequencies with very low loss of power, and with little or no interference.

Community antenna system Community antenna television (CATV) systems are those which employ a large master receiving antenna from which greatly enhanced signals are distributed, via cables, to various locations in a community.

Compatible color system One in which black-and-white signals are received normally by a color receiver, and vice versa.

Contrast The difference in brightness among various elements of a TV picture.

Control panel The master location in the control room where various selector switches for camera control are located; also may contain controls for other audio and video functions.

Control room The room in which the director and technical personnel control the production activity of a studio. They select the camera to be aired, and also shade the picture and perform other electronic functions.

Corner resolution It is inherent in a television system that the center of the picture is always sharper than the corners. The ability of an electronic system to resolve detail in these corners is called corner resolution.

Cover shot Same as establishing shot — a view of the entire set or viewing area to acquaint the viewer with the overall background.

Crop A type of camera framing which excludes some of the subject matter from the shot.

Crossfade A type of transition which involves a very fast fade to black and a fade back into the next scene. This term is more often used to signify the crossing of two sound channels, with a brief separation of silence.

Cross shot A condition in which two cameras are shooting from opposite sides of the set, with first one camera being taken and then the other.

CRT An abbreviation for cathode ray tube. It is the picture tube of a television system.

Cue The signal, either voiced or made visually, for action to begin.

Cue card A card placed on the set where the performer can see it easily. It contains script information, lyrics of songs, or other information troublesome to remember. Cue cards are also facetiously called "idiot cards."

Cut When used as a command by the director during production, it means to stop all action on the set. It also, however, refers to the fastest method of getting from one scene to another, by instantly replacing one picture with another.

Cutoff The part of the motion picture frame or transmitted picture which is not visible when seen on television receivers. A television tube tapers slightly; thus, the only part used for transmission is the flat surface. This usage avoids image distortion along the outer edges.

Decoder A part of television receiving equipment which receives the encoded signal and sorts out those components which are green, red, or blue. The processed signals then modulate the intensity of the color television receiver's three electron beams, thus allowing the duplication of the colors in the original subject matter. See *encoder*.

Definition A term that also means resolution. It refers to the ability of a television system to focus sharply and with detail.

Degauss A term that refers to the process of demagnetizing tape for reuse.

Demodulator output The device in videotape recorders (containing the postemphasis circuit) which separates tape sync from tape signal.

It provides line drivers which feed unprocessed video to the processing amplifier and to monitoring circuits. See *processor.*

Depth of field That portion of the picture, front to back, that is properly in focus.

Directional A term to describe the characteristic of a microphone to pick up sound from a certain angle. Directional implies only one direction. Omnidirectional refers to a microphone with the ability to pick up sound from all directions.

Director The person in the television production who has charge of all matters of action and composition in the production, and who supervises the efforts of technicians, cameramen, etc.

Dissolve A gradual transition accomplished by the simultaneous fading out of one picture and the fading in of another.

Distortion A condition during transmission or amplification in which the received signal wave form departs from that of the original transmitted wave form.

Dolly A dolly refers both to a type of camera mount, and to the action of moving the camera towards (dolly in) or away from (dolly out) the subject with the dolly (camera).

Down and under (or out) A cue to the audio personnel to reduce the sound level under dialogue, or to fade that sound out altogether.

Dropout A loss of video signal that is caused either by faulty tape-to-head separation or by insufficient coating of magnetic material on the tape.

Dropout compensator A device for videotape recorders which replaces missing video information.

Dub (also dupe) 1. The process of making copies of recorded information, either video or audio. 2. A copy.

Electron beam A linear batch of focused electrons which strike the fluorescent screen or chemical coating.

Encoder A device through which a television color signal is processed electronically. It combines the three color signals into one signal; thus, after transmission, the color components of the scene can duplicate the colors of the original subject matter and can be restored to their proper relationships.

Erasure The neutralization of the recorded magnetic pattern on video tape. It is accomplished by placing the tape in a strong magnetic field. The process is called *degaussing.*

Establishing shot A wide view of the entire scene so that the audience can identify the various components of the set, and also the placement of the performers.

Extreme close-up (ECU) A very tight view of the performer or the subject matter. A close-up (CU) might consist of a picture of a performer's head. An extreme close-up (ECU) would show only his eyes or some other portion of the head.

Fade A fade-in, or a fade-out, refers to the gradual increase (or decrease) in the video level of the picture. In other words, the picture emerges from black, or recedes into black.

Feedback A loud squeal emitted from a speaker when a microphone is placed in the wrong relationship to it, normally in front of the speaker.

Field One complete trace of the scanning beam across the face of the picture. Two fields are required to make one "frame," the scanning beam alternatively scanning the odd and the even lines.

Field frequency The electronic scanning rate of a complete field, or one half of a composite picture. The field frequency rate for the United States is 60 fields per second and for Europe, 50 fields per second.

Field pulse An editor's splicing guide. Magnetized impulses, which rest along the edge of a length of videotape.

Fill light The general illumination in a studio situation.

Film chain The equipment used so that 35mm slides and motion picture film can be used on television. It includes an optical system for focusing the pictures on the pickup tube of a television camera. A multiplexer is a film chain that includes two or three projectors which, through the use of mirrors, supply a picture to a single TV camera. A uniplexer utilizes only one projector.

First generation The original or master video tape. The first copy from this material becomes "second generation."

Flat The backdrops used on a TV studio set. They are made of hardboard, plastic, paper, and many other materials that are chosen depending on the need for one-time or multiple use.

Flip stand This consists of a system in which cards containing titles, artwork, or photographs are mounted in a ring binder. The material can be shown in sequence by flipping the cards forward (in front of the camera).

Flood Lights used for overall illumination. Also called scoops.

Floor manager A person who works on the studio floor during production and follows the director's orders with regard to the direction of on-camera talent. He or she also prepares the set for production efforts.

Focusing Adjusting the camera lens so that an acceptable sharp picture results.

Frame One complete television picture, made up of two complete fields. Frames are scanned at the rate of 30 frames per second.

Frame pulse Also called by some "edit pulse," this is a pulse superimposed on the control track signal to identify the longitudinal position of a video track containing a vertical synchronizing pulse. It is used as an aid in editing and the synchronization of some recorders.

Frequency response A sound system will record and reproduce accurately a certain range of tones or musical notes, known as the frequency response. Frequencies from approximately 20 to 16,000 cycles per second are capable of being heard by the human ear.

Funnel In a television studio, these metal tubes, also known as snoots, are stationed in front of spotlights and control the beam of light which originates from the source.

Gain The controlling of volume during the recording and reproduction processes. Gain controls are necessary to keep the amplitude of the sound signal within the limits of the equipment, within which, the equipment will reproduce faithfully.

Generation Original material. First-generation material, is referred to as a master. Duplicate material, a dub of a master, etc., is referred to as second-generation (or third-, fourth-, fifth-generation, etc.).

Gobo A two-dimensional piece of artwork placed between the camera and a set and used to create a graphic illusion.

Halo A condition caused usually by a bright object on a dark background. It causes a flare of light. (Women's jewelry can cause this effect.)

Head The signal pickup and recording portion of the video tape recorder's rotating drum.

Head alignment On a tape recorder, the reproduce and record heads must be positioned so that their gaps are mutually perpendicular and parallel to the path of travel of the tape. This positioning is known as head alignment.

Head channel The signal path unique to each magnetic head.

Head clogging The accumulation of debris (usually oxide particles) on the head, the usual result of which is a loss of signal during playback, and degradation or failure to record during the record mode.

Head room The space between the top of the performer's head and the top of the camera picture. The camera operator must allow some extra space so that the top of the performer's head is not cropped during dubbing.

High key The arrangement of television studio set lighting, which allows gradations between white and intermediate shades of grey to be produced in the print material. High-key scenes are scenes with almost no dark areas, and which are brightly lit.

Highlights The maximum brightness of the picture, or the areas which have the greatest illumination.

Hot A condition in which too much light is falling on a performer, an object, or a portion of the set.

Idiot cards Cue cards.

Impedance The rating of the input or output of an electronic component. This electrical rating is regarded as either "high" or "low" and it is important that impedances match when two components are interconnected.

Interference Any extraneous energy which tends to interfere with the reception of a transmitted signal.

Interlace The electronic process of scanning alternate lines (fields) of a TV picture so that flicker is minimized.

Iris An adjustable mask in a camera lens that limits the amount of light entering the camera. Depth of field is a function of the amount of light entering the lens — the smaller the iris opening, the greater the depth of field.

Jitter The short, vertical, or horizontal movements of a picture, which may appear as individual scanning lines or which may disrupt the entire field of view. This lack of synchronization will persist until the controls are adjusted.

Key light Highlight illumination on a set. It is used to accent the overall background illumination. Spotlights are most often used for this purpose.

Kinescope recording Also called a *kine*, this is simply a motion picture film, made from a television monitor.

Lag A condition in which an image on the television screen "lags" for a few frames behind the actual movement.

Lap dissolve See dissolve.

Lead-in The opening portion of a program, often with voice-over narration or music.

Level The average intensity of a video or audio source, as indicated on a VU (volume unit) meter.

Limbo A type of lighting in which the background seems to stretch into infinity, or in which the subject is presented against a black background.

Line The main program line or video source that feeds, for example, a line monitor.

Line frequency The number of horizontal scans per second, or about 15,750 per second. It is found by multiplying the number of frames (30 per second) by the number of lines per frame (525).

Live This refers to a camera or microphone that is on and being used for telecasting or recording. A "live" show is one that is telecast directly, not taped and shown later.

Long shot A wide shot on a small studio set. It usually includes full-length views of the performers.

Low key The arrangement of television studio set lighting, which allows gradation between black and the middle shades of grey. Low key scenes are those which are dark.

Magnetic tape A tape consisting of a flexible base material usually coated on one side with a thin magnetizable layer.

Mask Material used to cover some part of the set or picture that the director does not wish to have included.

Master The original video recording made during a telecast.

Microwave One of the methods of transmitting television signals through the air. It involves the use of a directional, line-of-sight system from originating to receiving stations.

Mil A unit of measure which is used to measure the thickness of recording tape. A mil equals 1/1000 inch.

Monitor A television set that accepts direct cable feeds from TV cameras or broadcast signals. *To* monitor means to observe picture shading and other factors.

Monochrome Having shades of only one color. Refers to black-and-white television.

Multiplexer An optical device for using a single television camera in conjunction with one or more motion picture or slide projectors.

Narrator The performer, usually off-camera, who delivers lines of copy in a televised program. If he or she is entirely off-camera, then that person's words are designated as "voice-over."

Negative image A condition in which the polarity of the picture is exactly opposite that of normal, and the resulting picture appears black in white areas and vice versa.

Noise Random electrical interference that produces a grainy, salt-and-pepper effect on the screen. Heavy noise is called "snow."

Output A connector which allows the transfer of signal output from one piece of equipment to another. It is used both in sound recording and videotaping.

Overloading In excess of a piece of equipment's frequency response; that is, sound recording and reproduction at volumes which go beyond the instrument's capacity to record or reproduce sound without distortion.

Pairing Problems in which the interlace of scanning lines of alternate fields fall (in pairs) on top of each other instead of between each other.

Pan A movement of the camera lens to the left or right.

Passing shot The stationary positioning of a camera, while the subject moves past it; or, the stationary positioning of the subject, while the camera scans past.

Pedestal A wheeled camera base. Some pedestals have the capability of up-and-down motion.

Picture tube The cathode ray tube of a television monitor or set. It creates an image by variations of an electron beam that scans the coated surface of the interior of the vacuum tube.

Pole tips Those parts of the video head which protrude radially beyond the rim of the head wheel and form the magnetic path to and from the tape.

Potentiometer Also called a "pot," this is a volume control dial or rheostat.

Preamplifier An amplifier to raise signal levels from microphones, etc., to a point high enough to be usable by a power amplifier.

Print-through A video duplication process involving the transfer of the signals of a master tape to a slave tape by magnetic induction.

Raster The scanned or illuminated area of a television picture tube.

Rear-screen projection A process in which a picture (either slide or motion picture film) is projected from the back of a translucent screen. Performers may then work in front of this "background" as though it were real.

Reel A hub with flanges on which video tape or film is wound.

Remote pickups Televised material acquired at some distance from the studio by a mobile unit (and sent back by microwave) or by a permanently installed remote camera (can be sent by cable).

Resolution The ability of a video system to resolve or reproduce details of the subject matter. Low resolution is characterized by a blurred, soft picture in which small details cannot be seen. High resolution implies a picture that is sharp and clear.

RF Abbreviation of "radio frequency," a system of transmission using the radio spectrum to carry audio and video signals.

Rim light In taping, the outlining of the subject with a halo of light. This effect is achieved by having the source of light behind the subject.

Roll A condition in which the picture moves upward or downward on the monitor. It is caused by a lack of vertical synchronization.

Running time The estimated (and later the actual) time it takes for the presentation of a given show.

Scan, scanning The process of breaking down an image into a series of elements representing light values. These values are then expressed as a series of parallel horizontal lines traced from left to right from top to bottom.

Scanning line A television screen is filled with 525 narrow horizontal lines (U.S. standard), which are created by the electron spot in a cathode-ray tube. One such line is a scanning line. Highlights, halftones, and shadows in the original subject matter are recreated by variations in the intensity of the spot.

Scene A sequence in a television program. It may consist of one or more TV shots. The term may also refer to the setting itself; the "scene" in which the action is occurring.

Second generation A tape copy made from the master tape.

Shooting script A TV script which includes all copy and related camera shots, sound effects, lighting plans, music, and other pertinent production information.

Shot The pickup from a television camera.

Signal Information that has been transposed into electrical impulses. Signals are expressed in terms of strength (or voltage) and frequency (cycles per second).

Signal-to-noise ratio The ratio between the level of the recorded signal and the noise level induced by the recording process and recording equipment. The lower the noise, the better the quality of the picture and sound.

Smear A picture condition in which objects appear to extend horizontally beyond their normal boundaries or edges. A blurred or "smeared" effect is the result.

Sneak A gradual introduction of a sound, a dissolve, music, etc., so that its presence isn't immediately obvious to the audience.

Snoot A metal tube, also called a funnel, which is stationed in front of a spotlight to control the beam of light which originates from the source.

Snow Heavy random video noise.

Sock cue To introduce music very suddenly at peak volume. It is also called a "stinger," and can be used for emphasis and shock value.

Special effects Any system for creating an illusion on television: It can include the use of miniatures, models, gobos, and other electronic and mechanical effects.

Special-effects generator An electronic device that permits the changing of one image to another by means of a variety of wipes — horizontal, vertical, and from various patterns such as a circle, a square, a diamond, etc.

Spindle The shaft on a tape recorder that passes through the drive hole in the tape reel.

Splice A butt-joint between two pieces of tape (or film) held together by means of a strip of self-adhesive.

Splicing tape A special material used to splice magnetic tape.

Split screen An electronic effect in which two different pictures appear side by side on the same screen. It can be accomplished with the use of a special-effects generator.

Stand by The cue which indicates to performers and crew that taping is about to begin. It is preceded by the request of "Quiet in the studio, please."

Strike To dismantle a set or take away props.

Superimpose To overlap the image from one source over the image from another source. It is also called a "super."

Sweep The repeated vertical and horizontal movement of an electron beam across the cathode-ray tube's fluorescent screen or chemical coating.

Switcher 1. A control that enables switching from one camera (or video source) to another. 2. The person who operates the switching equipment.

Sync An abbreviation for the word "synchronization." It refers to the process of producing synchronizing signals, or timing pulses, which keep various television circuits in phase with one another.

Talent The performers who appear in a television production.

Tally light The red light on a TV camera, when lit, that indicates it is the taking camera.

Tape guides Grooved pins on either side of the recording head assembly that serve to position the magnetic tape so that it feeds properly into the head in either the record or playback mode.

Tearing A condition of the TV picture in which groups of horizontal lines are displaced in an irregular manner, usually because of a lack of horizontal synchronization or too high a video level.

Technical director (TD) The person in charge of the electronic portion of the recording session. He or she may also do switching for the director.

Termination The resistance value placed across the receiving end of a coaxial cable (video signals are terminated by 75 ohms).

Test pattern A specially made pattern of lines or circles that is used for setting focus, linearity, registration, balancing, and otherwise lining up a television camera.

Tilt A vertical camera movement, tilting either up or down (in contrast to *panning* left or right).

Transfer To record from video tape to film, or the other way around.

Truck Camera movement, on a pedestal, in which the camera is moved either left or right in travel that is parallel to the set. (It contrasts with dollying in or out.)

Turret A camera mounting that has accommodation for up to four lenses. The mount rotates to bring successive lenses into position in front of the camera.

Vertical Sweep The vertical movement of the scanning beam when the picture image is being recreated over the television screen's 525 (U.S. Standard) scanning lines.

Video Pertaining to the picture portion of the television signal.

Video tape Magnetic recording tape intended for recording and playback of television signals.

Viewfinder A small TV monitor built into the camera that enables the cameraman to see precisely what is being aired or recorded.

Voice over The words spoken by an off-camera narrator.

VTR Video tape recorder.

Wipe A transition in which a new scene gradually replaces the old scene. In a typical horizontal wipe, one picture seems to advance and covers the other picture.

Wow Variations in tape speed which cause subsequent distortion of audio signals.

Zoom A multipurpose lens that is continuously variable (and in focus) everywhere from a wide angle to a telephoto position. To "zoom in" means to go for a shot in which the subject of interest slowly becomes larger in the picture. To "zoom out" means to pull back to a much wider shot.

ASSOCIATIONS, MANUFACTURERS, PRODUCTION FACILITIES, AND SUPPLIERS

Associations

American Society for Training and Development (ASTD)
313 Price Place
Madison, Wisconsin 53705

Association for Educational Communications and Technology (AECT)
1201 16th St., N.W.
Washington, D.C. 20036

Industrial Audio-visual Assoc.
P.O. Box 656 Downtown Sta.
Chicago, Illinois 60690

National Association of Educational Broadcasters (NAEB)
1346 Connecticut Avenue
Washington, D.C. 30006

National Audio-visual Association (NAVA)
3150 Spring Street
Fairfax, Va. 22030

International Industrial Television Association (ITVA)
P.O. Box 297
Summit, N.J. 07901

Society of Motion Picture and TV Engineers (SMPTE)
9 E. 41st Street
New York, New York 10017

Audio equipment

Altec (LTV Ling Altec, Inc.)
1515 S. Manchester Avenue
Anaheim, Calif. 92803

Ampex Corporation
401 Broadway
Redwood City, Calif. 94063

Bell and Howell Co.
7100 McCormick Road
Chicago, Illinois 60645

Collins Radio Company
1200 No. Alma Road
Dallas, Texas 75207

Concord Electronics Corp.
1935 Armacost Avenue
Los Angeles, Calif. 90025

Hamilton Electronics Corp.
2726 W. Pratt Blvd.
Chicago, Illinois 60645

Panasonic (Matsushita Electric Corp. of America)
Pan Am Bldg., 200 Park Avenue
New York, N.Y. 10017

Philips Broadcast Equipment Corp. (North American Philips Corp.)
1 Philips Parkway
Montvale, N.J. 07645

RCA Corporation
30 Rockefeller Plaza
New York, N.Y. 10020

Sony/Superscope
8150 Vineland
Sun Valley, Calif. 91352

Sparta Electronics Corp.
5851 Florin-Parkins Rd.
Sacramento, Calif. 95828

3M Company (Wollensak—Mincom Division)
3M Center
St. Paul, Minn. 55101

Vital Industries, Inc.
3614 S.W. Archer Rd.
Gainesville, Fla. 32601

Control consoles

Ampex Corporation
401 Broadway
Redwood City, Calif. 94063

Audiotronics Corporation
7428 Bellaire Ave.
North Hollywood, Calif. 91603

Cohu Electronics, Inc.
Box 623
San Diego, Calif. 92112

Dage (Visual Educom, Inc.)
U.S. Highway 12 East
Michigan City, Indiana 46360

GBC Closed Circuit TV Corp.
74 Fifth Ave.
New York, New York 10011

NASCO Television Systems
947 Janesville Ave.
Fort Atkinson, Wisc. 53530

Philips Broadcast Equipment Corp. (North American Philips Corp.)
1 Philips Parkway
Montvale, N.J. 07645

RCA Corporation
30 Rockefeller Plaza
New York, N.Y. 10020

Shintron Company, Inc.
144 Rogers St.
Cambridge, Mass. 02142

Telemation, Inc.
2195 South 3600 West
Salt Lake City, Utah 84119

Film-chain equipment (telecine systems)

Buhl Optical Company
1009 Beech Ave.
Pittsburgh, Pa. 15233

Kalart Victor Corp.
Hultenius St.
Plainville, Conn. 06062

L-W International
6416 Variel Ave.
Woodland Hills, Calif. 91364

Magnavox Video Systems
2829 Maricopa St.
Torrence, Calif. 90503

Philips Broadcast Equipment Corp. (North American Philips Corp.)
1 Philips Parkway
Montvale, N.J. 07645

RCA Corporation
30 Rockefeller Plaza
New York, N.Y. 10020

Riker Communications, Inc.
142 Central Ave.
Clark, N.J. 07066

Spindler & Sauppé
13034 Saticoy St.
No. Hollywood, Calif. 91605

Tele-Cine, Inc.
294 East Shore Drive
Massapequa, N.Y. 11758

Telemation, Inc.
2195 South 3600 West
Salt Lake City, Utah 84119

Lenses

Canon U.S.A., Inc.
64-10 Queens Blvd.
Woodside, N.Y. 11377

Cohu Electronics, Inc.
Box 623
San Diego, Calif. 92112

Panasonic (Matsushita Electric Corp. of America)
Pan Am Bldg., 200 Park Ave.
New York, N.Y. 10017

Tele-Cine, Inc.
294 East Shore Drive
Massapequa, N.Y. 11758

Zoomar, Inc.
55 Sea Cliff Ave.
Glen Cove, N.Y. 11542

Lighting equipment and controls

Bardwell & McAlister, Inc.
12164 Sherman Way
No. Hollywood, Calif. 91605

Century Strand, Inc.
3411 W. El Segundo Blvd.
Hawthorne, Calif. 90250

Colortran (Berkey Colortran Corp.)
1015 Chestnut Street
Burbank, Calif. 91502

GTE Sylvania
Lighting Center
Danvers, Mass. 01923

Kliegl Brothers
32-32 48th Ave.
Long Island City, N.Y. 11101

Mole-Richardson Co.
937 N. Sycamore Ave.
Hollywood, Calif. 90038

Skirpan Lighting Control Corp.
41-43 24th St.
Long Island City, N.Y. 11101

Microphone accessories (mixers, connectors, cables, etc.)

Altec (LTV Ling Altec, Inc.)
1515 S. Manchester Ave.
Anaheim, Calif. 92803

Collins Radio Company
1200 No. Alma Road
Dallas, Texas 75207

Shure Brothers, Inc.
222 Hartrey Ave.
Evanston, Ill. 60204

Sony Corporation of America
47-47 Van Dam Street
Long Island City, N.Y. 11101

Television & Computer Corp.
2385 Beryllium Road
Scotch Plains, N.J. 07076

Microphones

Conrac (Turner Division of Conrac Corp.)
909 17th Street, N.E.
Cedar Rapids, Iowa 52402

Electro-Voice, Inc.
600 Cecil Street
Buchanan, Mich. 49107

Panasonic (Matsushita Electric Corp. of America)
Pan Am Bldg., 200 Park Ave.
New York, N.Y. 10017

Shure Brothers, Inc.
222 Hartrey Ave.
Evanston, Ill. 60204

Sony Corporation of America
47-47 Van Dam Street
Long Island City, N.Y. 11101

Special-effects generators

Alma Engineering, Inc.
7990 Dagget St.
San Diego, Calif. 92111

Ball Brothers Research Corp. (Miratel Division)
1633 Terrace Drive
St. Paul, Minn. 55113

Dynasciences Video Products
Township Line Rd.
Blue Bell, Pa. 19422

The Grass Valley Group, Inc.
P.O. Box 1114
Grass Valley, California 95945

Riker Communications, Inc.
142 Central Avenue
Clark, New Jersey 07066

Shintron Company, Inc.
144 Rogers Street
Cambridge, Mass. 02142

Sony Corporation of America
47-47 Van Dam Street
Long Island City, N.Y. 11101

Switcher/faders

Alma Engineering, Inc.
7990 Dagget St.
San Diego, Calif. 92111

Ball Brothers Research Corp. (Miratel Division)
1633 Terrace Drive
St. Paul, Minn. 55113

Cohu Electronics, Inc.
Box 623
San Diego, Calif. 92112

Dage (Visual Educom, Inc.)
U.S. Highway 12 East
Michigan City, Indiana 46360

Dynair Electronics, Inc.
6360 Federal Boulevard
San Diego, Calif. 92114

Dynasciences Video Products
Township Line Rd.
Blue Bell, Pa. 19422

The Grass Valley Group, Inc.
P.O. Box 1114
Grass Valley, Calif. 95945

Riker Communications, Inc.
142 Central Ave.
Clark, N.J. 07066

Shintron Company, Inc.
144 Rogers Street
Cambridge, Mass. 02142

Telemation, Inc.
2195 South 3600 West
Salt Lake City, Utah 84119

Vital Industries, Inc.
3614 S.W. Archer Rd.
Gainesville, Fla. 32601

Synchronizing amplifiers

Ball Brothers Research Corp. (Miratel Division)
1633 Terrace Drive
St. Paul, Minn. 55113

Shintron Company, Inc.
144 Rogers Street
Cambridge, Mass. 02142

Synchronizing generators

Ball Brothers Research Corp. (Miratel Division)
1633 Terrace Drive
St. Paul, Minn. 55113

Cohu Electronics, Inc.
Box 623
San Diego, Calif. 92112

Dage Television
U.S. Highway 12 East
Michigan City, Indiana 46360

Diamond Power Specialty Corp.
P.O. Box 415
Lancaster, Ohio 43130

The Grass Valley Group, Inc.
P.O. Box 1114
Grass Valley, California 95945

Panasonic (Matsushita Electric Corp. of America)
Pan Am Bldg., 200 Park Avenue
New York, N.Y. 10017

RCA Corporation
30 Rockefeller Plaza
New York, N.Y. 10020

Riker Communications, Inc.
142 Central Avenue
Clark, N.J. 07066

Shintron Company, Inc.
144 Rogers Street
Cambridge, Mass. 02142

Telemation, Inc.
2195 South 3600 West
Salt Lake City, Utah 84119

Tele-prompters

Q-TV Sales & Distributing Corp.
342 W. 40th Street
New York, N.Y. 10018

Telesync Corporation
20 Insley Street
Demarest, N.J. 07627

Television monitors

Ampex Corporation
401 Broadway
Redwood City, Calif. 94063

Audiotronics Corporation
7428 Bellaire Avenue
North Hollywood, Calif. 91603

Ball Brothers Research Corp. (Miratel Division)
1633 Terrace Drive
St. Paul, Minn. 55113

Cohu Electronics, Inc.
Box 623
San Diego, Calif. 92112

Concord Electronics Corp.
1935 Armacost Avenue
Los Angeles, Calif. 90025

Conrac Corporation
600 N. Rimsdale Ave.
Covina, Calif. 91722

Craig Corporation
921 West Artesia Blvd.
Compton, Calif. 90220

GBC Closed Circuit TV Corp.
74 Fifth Avenue
New York, N.Y. 10011

Javelin Electronics Div.
6357 Arizona Circle
Los Angeles, Calif. 90045

JVC Industries, Inc.
50-35 56th Rd.
Maspeth, N.Y. 11378

The Magnavox Company
Video Systems Dept.
Fort Wayne, Ind. 46804

Panasonic (Matsushita Electric Corp. of America)
Pan Am Bldg., 200 Park Ave.
New York, N.Y. 10017

RCA Corporation
30 Rockefeller Plaza
New York, N.Y. 10020

Riker Communications, Inc.
142 Central Avenue
Clark, N.J. 07066

SC Electronics, Inc.
530 5th Ave., N.W.
New Brighton, Minn. 55112

Shibaden Corporation of America
58-25 Brooklyn-Queens Exwy.
Woodside, N.Y. 11377

Sony Corp. of America
47-47 Van Dam St.
Long Island City, N.Y. 11101

World Video, Inc.
P.O. Box 117
Boyertown, Pa. 19512

Time base correctors

Ampex Corporation
401 Broadway
Redwood City, Calif. 94063

Consolidated Video Systems
300 Edward Ave.
Santa Clara, Calif. 95050

International Video Corp.
990 Almanor Ave.
Sunnyvale, Calif. 94086

Television Microtime, Inc.
1280 Blue Hills Ave.
Bloomfield, Conn. 06002

Tripods, dollies and heads

Davis and Sanford Co., Inc.
24 Pleasant St.
New Rochelle, N.Y. 10802

Listec Television Equipment Corp.
35 Cain Drive
Plainview, N.Y. 11803

Power-Optics, Inc.
1055 W. Germantown Pike
Fairview Village, Penna. 19409

Quick-Set, Inc.
3650 Woodhead Drive
Northbrook, Ill. 60062

Riker Communications, Inc.
142 Central Avenue
Clark, N.J. 07066

Technology Incorporated (HF Photo Systems Div.)
11801 W. Olympic Blvd.
Los Angeles, Calif. 90064

Video cameras

Ampex Corporation
401 Broadway
Redwood City, Calif. 94063

Audiotronics Corporation
7428 Bellaire Ave.
North Hollywood, Calif. 91603

Cohu Electronics, Inc.
Box 623
San Diego, Calif. 92112

Commercial Electronics, Inc.
880 Maude Ave.
Mt. View, Calif. 94040

Concord Electronics Corp.
1935 Armacost Avenue
Los Angeles, Calif. 90025

Craig Corporation
921 West Artesia Blvd.
Compton, Calif. 90220

Dage (Visual Educom, Inc.)
U.S. Highway 12 East
Michigan City, Indiana 46360

Diamond Power Specialty Corp.
P.O. Box 415
Lancaster, Ohio 43130

GBC Closed Circuit TV Corp.
74 Fifth Ave.
New York, N.Y. 10011

Hitachi Shibaden Corp. of America
58-25 Brooklyn-Queens Exwy.
Woodside, N.Y. 11377

International Video Corp.
675 Almanor Ave.
Sunnyvale, Calif. 94086

JVC Industries, Inc.
50-35 56th Road
Maspeth, N.Y. 11378

Magnavox Video Systems
Video Systems Dept.
Fort Wayne, Ind. 46804

Panasonic (Matsushita Electric Corp. of America)
Pan Am Bldg., 200 Park Avenue
New York, N.Y. 10017

Philips Broadcast Equipment Corp. (North American Philips Corp.)
1 Philips Parkway
Montvale, N.J. 97645

RCA Corporation
30 Rockefeller Plaza
New York, N.Y. 10020

Riker Communications, Inc.
142 Central Avenue
Clark, N.J. 07066

Sharp Electronics Corp.
10 Keystone Place
Paramus, N.J. 07652

Sony Corporation of America
47-47 Van Dam St.
Long Island City, N.Y. 11101

Telemation, Inc.
2195 South 3600 West
Salt Lake City, Utah 84119

Video Projectors

Advent Corporation
195 Albany Street
Cambridge, Mass. 02139

General Electric Company
Electronics Park, 6-206
Syracuse, N.Y. 13201

Sony Corporation of America
47-47 Van Dam St.
Long Island City, N.Y. 11101

Video tape (blank)

Ampex Corporation
401 Broadway
Redwood City, California 94063

Dale Industries
108 Vanowen St.
North Hollywood, Calif. 91605

Irish Magnetic Recording Tape
270-78 Newtown Road
Plainview, N.Y. 11803

Memorex Corporation
1200 Memorex Drive
Santa Clara, Calif. 95052

Panasonic (Matsushita Electric Corp. of America)
Pan Am Bldg., 200 Park Avenue
New York, N.Y. 10017

Sony Corporation of America
47-47 Van Dam Street
Long Island City, N.Y. 11101

3M Company
3M Center
St. Paul, Minn. 55101

Wabash Tape Corp.
P.O. Box 128
Huntley, Illinois 60142

Video tape recorders (VTR) 1" and 1/2"

Ampex Corporation
401 Broadway
Redwood City, Calif. 94063

Audiotronics Corporation
7428 Bellaire Avenue
North Hollywood, Calif. 91603

Concord Electronics Corp.
1235 Armacost Avenue
Los Angeles, Calif. 90025

Craig Corp.
921 W. Artesia Blvd.
Compton, Calif. 90220

Diamond Power Specialty Corp.
P.O. Box 415
Lancaster, Ohio 43130

GBC Closed Circuit TV Corp.
74 Fifth Ave.
New York, N.Y. 10011

International Video Corp.
675 Almanor Ave.
Sunnyvale, Calif. 94086

JVC Industries, Inc.
50-35 56th Rd.
Maspeth, N.Y. 11378

Javelin Electronics Co.
6357 Arizona Circle
Los Angeles, Calif. 90045

Panasonic (Matsushita Electric Corp. of America)
Pan Am Bldg., 200 Park Avenue
New York, N.Y. 10017

RCA Corporation
30 Rockefeller Plaza
New York, N.Y. 10020

Hitachi Shibaden Corp. of America
58-25 Brooklyn-Queens Exwy.
Woodside, N.Y. 11377

Sony Corporation of America
47-47 Van Dam St.
Long Island, N.Y. 11101

Video tape transfer and duplicating

AV Center
6116 N. Lincoln Avenue
Chicago, Ill. 60659

Byron Motion Pictures
65 K Street, N.E.
Washington, D.C. 20002

Dolphin Productions
305 E. 45th Street
New York, N.Y. 10017

E.U.E./Screen Gems
513 W. 54th Street
New York, N.Y. 10019

Golden West Videotape
5800 Sunset Blvd.
Hollywood, Calif. 90028

MPCS Communications
424 W. 49th Street
New York, N.Y. 10019

National Teleproductions Corp.
5261 N. Tacoma Avenue
Indianapolis, Ind. 46220

New England Video Services
501 Boylston St.
Boston, Mass. 02117

W.A. Palmer Films, Inc.
611 Howard Street
San Francisco, Calif. 94105

Reeves Production Services
304 E. 44th Street
New York, N.Y. 10017

Rombex Productions Corp.
245 W. 55th Street
New York, N.Y. 10019

S/T Videocassette Duplicating Corp.
220 E. 51st Street
New York, N.Y. 10022

Telematron Productions, Inc.
3200 W. West Lake Avenue
Glenview, Ill. 60025

Teletronics International
220 E. 51st Street
New York, N.Y. 10022

Television Associates, Inc.
1157 Saratoga Avenue
San Jose, Calif. 95129

Video Realities, Inc.
16 West Alder Street
Walla Walla, Wash. 99362

Video Software and Production Center, Inc.
165 Tuckahoe Road
Yonkers, N.Y. 10710

Video Trans, Inc.
211 E. Grand Avenue
Chicago, Ill. 60611

Videograf, Inc.
100 Cabot Street
Needham, Mass. 02194

Vidtronics, Inc.
342 Madison Avenue
New York, N.Y. 10017

Visual Information Systems
15 Columbus Circle
New York, N.Y. 10023

Windsor Electronics Systems Corp.
652 First Avenue
New York, N.Y. 10016

Production facilities (1-inch and 1/2-inch)

AV Center
6116 N. Lincoln Avenue
Chicago, Ill. 60659

Calvin Communications, Inc.
215 W. Pershing Road
Kansas City, Mo. 64106

E.U.E./Screen Gems
513 W. 54th Street
New York, N.Y. 10019

Eastern Airlines Television Center
Building 30, Miami International Airport
Miami, Fla. 33148

Golden West Videotape
5800 Sunset Blvd.
Hollywood, Calif. 90028

Lewron Television, Inc.
441 W. 53rd Street
New York, N.Y. 10019

MPCS Communications
424 W. 49th Street
New York, N.Y. 10019

National Teleproductions Corp.
5261 N. Tacoma Avenue
Indianapolis, Ind. 46220

New England Video Services
501 Boylston St.
Boston, Mass. 02117

Reeves Production Services
304 E. 44th Street
New York, N.Y. 10017

Rombex Productions Corp.
245 W. 55th Street
New York, N.Y. 10019

Telematron Productions, Inc.
3200 W. West Lake Avenue
Glenview, Ill. 60025

Teletronics International
220 E. 51st Street
New York, N.Y. 10022

Television Associates, Inc.
1157 Saratoga Avenue
San Jose, Calif. 95129

Video Realities, Inc.
16 West Alder Street
Walla Walla, Wash. 99362

Video Software and Production Center, Inc.
165 Tuckahoe Road
Yonkers, N.Y. 10710

Video Trans, Inc.
211 E. Grand Avenue
Chicago, Ill. 60611

Videograf, Inc.
100 Cabot Street
Needham, Mass. 02194

Vidtronics, Inc.
342 Madison Avenue
New York, N.Y. 10017

Visual Information Systems
15 Columbus Circle
New York, N.Y. 10023

INDEX